Helium Engineering Technologies

Liming Pu
Editor

Helium Engineering Technologies

From Source to Use

上海交通大學出版社
SHANGHAI JIAO TONG UNIVERSITY PRESS

Springer

Editor
Liming Pu
China Petroleum Engineering &
Construction Corporation
Beijing, China

ISBN 978-981-96-4085-0 ISBN 978-981-96-4086-7 (eBook)
https://doi.org/10.1007/978-981-96-4086-7

Jointly published with Shanghai Jiao Tong University Press
The print edition is not for sale in China (Mainland). Customers from China (Mainland) please order the print book from: Shanghai Jiao Tong University Press

This Springer imprint is published by the registered company Springer Nature Singapore Pte Ltd.
The registered company address is: 152 Beach Road, #21-01/04 Gateway East, Singapore 189721, Singapore

Preface

Helium, one of the noble gases, is represented by the symbol He. It is a colorless, odorless element with an atomic number of two on the periodic table and is the least dense gas after hydrogen. In the universe, helium is the second most abundant element after hydrogen, accounting for about 23% of the total mass of the cosmos. Nonetheless, it is extremely rare on Earth. As research delves deeper, the unique properties of helium are gradually coming to light, expanding its applications into increasingly diverse fields, playing an important role in promoting economic and high-tech industries. The predominant global method for helium production today involves extraction from natural gas, creating an integrated industrial chain that encompasses crude helium extraction, refining, liquefaction, storage, and transportation.

Drawing on past experience in helium extraction technology R&D, project construction, and plant operation, and referencing domestic and international literature, as well as relevant national and industry standards and technical manuals, the author team has compiled the book. The book consists of nine chapters, organized according to the main units involved in the helium extraction process. Chapter 1 introduces helium, detailing its properties, and typical applications. Chapter 2 discusses the distribution of helium resources and analyzes the supply-demand dynamics within the helium market. Chapter 3 covers the history of helium extraction from natural gas, the production of helium, helium storage and transportation, and the core technologies involved in these key stages of the helium supply chain. Chapter 4 introduces various crude helium extraction technologies, including common cryogenic technology, membrane separation technology, combined membrane separation and cryogenic technology, multi-product co-production technology, and multiple gas source technology. Chapter 5 discusses crude helium refining technologies, focusing on catalytic dehydrogenation and adsorptive dehydrogenation technologies, pressure swing adsorption (PSA) or temperature swing adsorption (TSA) technologies for the removal of nitrogen and hydrocarbon impurities, and cryogenic adsorption or cryogenic PSA technologies for the removal of neon and other impurities. Chapter 6 explores the extreme temperature conditions required for helium liquefaction, detailing the liquefaction process and introducing typical helium liquefiers at home and abroad. Chapter 7 covers various storage and transportation equipment for gaseous and liquid helium, applicable scenarios, and the underground storage method focused on helium storage reservoirs. Chapter 8 introduces the process equipment used for helium extraction, including the basic principles, operational characteristics, equipment selection, and application scenarios of commonly used equipment such as plate-fin heat exchangers, helium concentration columns, gas separation membranes, and helium turboexpanders. Chapter 9 discusses helium extraction plants, providing insights into typical plants such as the Amur Helium Extraction Plant, the Qatar Helium Extraction Plant, and the Rongxian Helium Extraction Plant. All resource data and engineering project information mentioned in the book are derived from publicly available sources, as noted in the text.

The development of helium extraction technologies in China is still in its developmental stages, with both the theoretical and practical aspects of these technologies undergoing continuous evolution. We invite readers to offer critiques and corrections for any shortcomings in the book.

Beijing, China Liming Pu

The original version of the book has been revised. A correction to this book can be found at https://doi.org/10.1007/978-981-96-4086-7_10

Contents

1 Properties and Applications of Helium 1
Liming Pu
1.1 What Is Helium 1
1.1.1 Discovery of Helium 2
1.1.2 Properties of Helium 4
1.1.3 Superfluid Helium 5
1.1.4 Origin of Helium 9
1.1.5 Research History of Helium 11
1.2 Typical Uses of Helium 12
1.2.1 Cryogenic Superconductivity 17
1.2.2 Aerospace 22
1.2.3 Optical Fiber and Semiconductor 24
1.2.4 Welding Protection 25
1.2.5 Chromatographic Analysis 26
1.2.6 Leak Detection Gas 27
1.2.7 Breathing Mixture 29
1.3 Quality Standards for Helium Products 31
1.3.1 Chinese Helium Product Standards 32
1.3.2 U.S. Helium Product Standards 32
1.4 Isotope Helium-3 (^3He) 33
1.4.1 Physical Properties of ^3He 35
1.4.2 Extraction of ^3He 36
1.4.3 Typical Applications of ^3He 38
References 41

2 Helium Resources and Market 43
Yingke Li and Yafei Zhang
2.1 Distribution of Helium Resources 44
2.1.1 Global Distribution of Helium Resources 44
2.1.2 Distribution of Helium Resources in China 45

2.2 Global Helium Market 47
2.2.1 Global Helium Production 47
2.2.2 Global Helium Demand 48
2.2.3 Global Helium Supply Chain 48
2.3 China's Helium Market 50
2.3.1 China's Helium Production Capacity 50
2.3.2 Helium Imports 50
2.3.3 Helium Consumption Sectors 53
2.4 Protection of Helium Resources 54
References 55

3 Industrial Chain of Helium Extraction from Natural Gas 57
Liming Pu and Yunqiang Chen
3.1 History of Extracting Helium from Natural Gas 58
3.1.1 History of Helium Extraction from Natural Gas in Other Countries 58
3.1.2 History of Extracting Helium from Natural Gas in China 60
3.2 Key Links in the Industrial Chain of Extracting Helium from Natural Gas 62
3.2.1 Crude Helium Extraction 64
3.2.2 Crude Helium Refining 67
3.2.3 Helium Liquefaction 68
3.2.4 Helium Storage and Transport 70
3.3 Helium-Related Standards 71
References 74

4 Crude Helium Extraction Technologies 75
Ke Wang, Yingjie Xiong, and Liming Pu
4.1 Cryogenic Technology 76
4.1.1 Flash Evaporation for Helium Extraction 77
4.1.2 Rectification for Helium Extraction 80
4.2 Membrane Separation Technology 84
4.2.1 Principles of Membrane Separation 85
4.2.2 Helium Extraction by Membrane Separation 88
4.3 Combined Membrane Separation and Cryogenic Technology 91
4.4 Multi-product Co-production Technology 92
4.4.1 Crude Helium Extraction Co-production with LNG 93
4.4.2 Crude Helium Extraction Co-production with Ethane 93
4.4.3 Crude Helium Extraction Co-production with Ethane and LNG 95
4.5 Multiple Gas Source Technology 96
4.5.1 BOG Crude Helium Extraction 96
4.5.2 Exhaust Gas Crude Helium Extraction 98
References 99

5 Crude Helium Refining Technologies 101
Xuan Zhou and Yingyong Zou
5.1 Catalytic Dehydrogenation Technology 101
5.1.1 Principle 102
5.1.2 Typical Process Flow 103
5.2 Adsorptive Dehydrogenation Technology 105
5.2.1 Principle 105
5.2.2 Typical Process Flow 106
5.3 Pressure Swing Adsorption Technology 107
5.3.1 Principle 107
5.3.2 Typical Process Flow 109
5.4 Temperature Swing Adsorption Technology 112
5.4.1 Principle 112
5.4.2 Typical Process Flow 113
5.5 Low-Temperature Adsorption Technology 113
5.5.1 Principle 114
5.5.2 Typical Process Flow 114
5.6 Low-Temperature Pressure Swing Adsorption Technology 116
5.6.1 Principle 116
5.6.2 Typical Process Flow 117
5.7 Technical Applications 117
References 120

6 Helium Liquefaction Technologies 121
Miao Yang and Limin Qiu
6.1 The Pursuit of "Absolute Zero" 122
6.1.1 Chlorine Liquefaction 122
6.1.2 Oxygen and Nitrogen Liquefaction 124
6.1.3 Hydrogen Liquefaction 126
6.1.4 Helium Liquefaction 128
6.2 Typical Helium Liquefaction Processes 130
6.2.1 Throttling Refrigeration Cycle 131
6.2.2 Expansion Refrigeration Cycle 133
6.2.3 Refrigeration Cycle with Gas Regenerative Refrigerator Precooling 134
6.3 Typical Helium Liquefaction Process Flows 135
6.4 Helium Liquefiers 136
6.4.1 Helium Liquefiers International 136
6.4.2 Helium Liquefiers in China 140
References 145

7 Helium Storage and Transport 147
Runlin Sun, Yingke Li, Guangzhao Li, and Hao Cui
7.1 Gaseous Helium Storage and Transport 148
7.1.1 Storage Tanks 148
7.1.2 Tube Bundle 150

7.1.3 Gas Cylinder ... 153
7.1.4 Gas Holder ... 154
7.1.5 Helium Storage Reservoir ... 154
7.1.6 Gaseous Helium Filling ... 159
7.2 Liquid Helium Storage and Transport ... 161
7.2.1 Liquid Helium Tank Container ... 161
7.2.2 Liquid Helium Dewar Flask ... 164
7.2.3 Liquid Helium Loading ... 167
7.3 Selection of Storage and Transport Methods ... 168
7.3.1 Selection of Gaseous Helium Storage and Transportation Equipment ... 168
7.3.2 Selection of Liquid Helium Storage and Transportation Equipment ... 168
7.3.3 Comparison of Gaseous Helium and Liquid Helium Storage and Transport ... 169
References ... 171

8 Equipment for Helium Extraction from Natural Gas ... 173
Zhiming Jiang and Yingjie Xiong
8.1 Plate-Fin Heat Exchanger ... 173
8.1.1 Structure ... 174
8.1.2 Features ... 174
8.1.3 Applications ... 176
8.2 Helium Concentration Column ... 176
8.2.1 Structure ... 177
8.2.2 Features ... 179
8.2.3 Applications ... 180
8.3 Gas Separation Membrane ... 180
8.3.1 Types of Membrane Modules ... 181
8.3.2 Membrane Module Materials ... 183
8.3.3 Applications ... 187
8.4 Compressor ... 188
8.4.1 Centrifugal Compressor ... 189
8.4.2 Reciprocating Compressor ... 190
8.4.3 Screw Compressor ... 191
8.4.4 Diaphragm Compressor ... 192
8.4.5 Applications ... 194
8.5 Helium Turboexpander ... 195
8.5.1 Structure ... 195
8.5.2 Features ... 196
8.5.3 Applications ... 197
References ... 198

9 Helium Extraction Plants from Natural Gas 201
Haiyang Long and Jianhua Tan
9.1 Typical Helium Extraction Plants out of China 201
9.1.1 Russia's Amur Gas Processing Plant 201
9.1.2 Qatar's Helium Extraction Plants 203
9.1.3 Darwin Helium Extraction Plant in Australia 206
9.1.4 Mankota Helium Extraction Plant in Canada 207
9.1.5 Odolanów Helium Extraction Plant in Poland 208
9.2 Typical Helium Extraction Plants in China 210
9.2.1 Rongxian Helium Extraction Plant of Chengdu Natural Gas Chemical Complex 210
9.2.2 BOG Helium Extraction Plant of Inner Mongolia Xingsheng Natural Gas Co., Ltd. 212
9.2.3 BOG Helium Extraction Plant of Yulin Yuanheng Energy Co., Ltd. 215
References 217

Correction to: Helium Engineering Technologies C1
Liming Pu

About the Editor

Liming Pu born in 1982 in Nanchong, Sichuan, received a Ph.D. from Zhejiang University. He is Senior Engineer and Certified Chemical Engineer in China. Pu currently serves as President of the Oil and Gas Field Surface Engineering Technology Branch of China Petroleum Engineering CORP., LTD. Engineering Technology R&D Institute, Chief Engineer at China Petroleum Engineering & Construction Corporation Southwest Company and Deputy Chief Engineer at the National Energy R&D Center for LNG Technology. He also serves as Deputy Editor-in-Chief of the journal *Natural Gas and Oil*. He has dedicated his career to technology R&D and engineering applications in cryogenics such as natural gas liquefaction, natural gas condensate recovery, and helium extraction from natural gas. Pu has filed over 40 invention patents, and has published more than ten academic papers on top energy-related academic journals.

Chapter 1
Properties and Applications of Helium

Liming Pu

Helium (He) and hydrogen (H) are the most abundant elements in the universe and the main components of the Sun, where hydrogen constitutes approximately 74.9% of the total mass and helium accounts for about 23.8%. At the Sun's core, hydrogen atoms are transformed to helium atoms through nuclear fusion, unleashing a torrent of energy.

Helium was first discovered in the Sun in 1868 and later identified on Earth in 1895. Given its unique physical and chemical properties, helium has been a research hotspot for over a century, leading to the emergence of new phenomena, theories, and achievements. Research on helium has not only expanded our understanding of the universe but also propelled scientific and technological advancements. Once the provider of buoyancy for balloons and airships, helium now plays a pivotal role in a wide spectrum of high-tech manufacturing sectors such as aerospace, healthcare, national defense, and superconductivity. It has become an indispensable strategic resource fueling progress in both defense and high-tech industries. This chapter introduces the properties and applications of helium.

1.1 What Is Helium

Helium (He) is the second element in the periodic table, with an atomic number of 2 and a CAS number of 7440-59-7. In normal temperature and pressure, helium is a colorless, odorless gas that exhibits no chemical reactivity. Known for its extreme stability and the lowest boiling point, helium is an inert and noble gas and the second-lightest element in nature, second only to hydrogen.

L. Pu (✉)
China Petroleum Engineering & Construction Corporation, Beijing, China
e-mail: puliming_sw@cnpc.com.cn

L. Pu (ed.), *Helium Engineering Technologies*,
https://doi.org/10.1007/978-981-96-4086-7_1

1.1.1 Discovery of Helium

On August 18, 1868, a total solar eclipse occurred in British India, garnering the attention of astronomers and physicists worldwide. Pierre Jules César Janssen, then director of the Astrophysical Observatory at Meudon, France, led a team to observe the eclipse in Guntur, Madras. Equipped with a telescope and a spectrometer, Janssen sought to study solar prominences—massive flares erupting from the Sun's surface. As the eclipse commenced, he trained his spectroscope towards the solar prominence, only to be met with an unexpected sight—a strikingly bright yellow spectral line of unknown origin. While reminiscent of sodium's spectral line, it deviated subtly from the expected pattern. Concerned about possible equipment issues, Pierre Jules César Janssen repeated his observations the following dawn. The results mirrored those from before, with identical spectral lines also detected at the solar limb [1]. He concluded that the spectral line with a wavelength of 587.49 nm was not associated with sodium and penned a letter detailing his findings to the French Academy of Sciences. Due to communication delays, his letter did not reach the Academy until October 24.

Coincidentally, on the very day Pierre Jules César Janssen's letter reached the Academy, a letter from Sir Joseph Norman Lockyer, then director of the Solar Physics Observatory of the Royal Society in England, also arrived. On October 20 of the same year, Joseph Norman Lockyer discovered the same yellow line while analyzing the solar spectrum in London and named it "D3". After laboratory comparisons, British chemist Sir Edward Frankland and Sir Joseph Norman Lockyer concluded that this spectral line belonged to an unknown element. They named it Helium, after the Greek word ἥλιος (Helios), the god of the Sun, earning helium the moniker "solar element".

Pierre Jules César Janssen and Sir Joseph Norman Lockyer are recognized as the discoverers of helium (Fig. 1.1). This marked the first discovery of helium outside Earth, uncovering its presence in the universe. For the next two decades, helium was only detected in the Sun, with no traces found on Earth.

Fig. 1.1 Pierre Jules César Janssen (left), Edward Frankland (center), Joseph Norman Lockyer (right)

Between 1888 and 1890, William Francis Hillebrand, a mineral chemist at the United States Geological Survey, observed that heating Cleveite in sulfuric acid released a gas. He collected this gas for experimentation, noting that it neither burned nor fueled combustion, did not turn limewater cloudy, and had no distinctive odor. Mistaking it for ordinary nitrogen, William Francis Hillebrand dismissed the significance and moved on. Helium is exceedingly rare on Earth, chemically inert, and resistant to bonding with other elements, making it challenging to detect. As a result, it eluded early chemists.

In 1894, John William Strutt, a professor of natural philosophy at the Royal Institution of Great Britain, and William Ramsay, a professor of chemistry at University College London, discovered the inert gas argon from the air, sparking interest in studying mysterious inert gases. In March 1895, chemistry professor Henry Myers, while reviewing literature, came across William Francis Hillebrand's report and speculated that the gas from uraninite might be argon, for which he engaged William Ramsay in verification. William Ramsay filled a discharge tube with the gas from Cleveite and, as a result, observed a yellow glow in the tube and a bright yellow spectral line in the spectroscope. This spectral line did not match argon's and was unlike any known element, reminding him of the "solar element"—helium—discovered 27 years earlier. For the sake of caution, William Ramsay enlisted the help of William Crookes, Britain's foremost spectroscopist, who confirmed the unknown gas was indeed helium. In March 1895, William Ramsay published a brief note in *Chemical News* on the discovery of helium on Earth and announced it at the annual meeting of the Royal Society of Chemistry later that year. Thus, helium, initially found in the celestial realm, was now confirmed also as an earthly element. Subsequently, scientists identified helium in the air, hot springs, and natural gas, albeit in very low concentrations.

The helium discussed above refers to the common isotope, ^{4}He, as the existence of isotopes was not yet known at the time. It was not until 1934 when Marcus Laurence Elwin Oliphant and others discovered ^{3}He that the concept of helium was distinguished [2, 3]. Today, we know helium belongs to Group 0 of the periodic table, with eight isotopes ranging from ^{3}He to ^{10}He. Among them, ^{3}He and ^{4}He are stable, while the others are unstable and radioactive.

Most of the ^{3}He on Earth is primordial helium captured from the solar system during Earth's formation, and it is extremely rare. Helium formed later on Earth, primarily in the form of ^{4}He, is produced through alpha decay of uranium (U) and thorium (Th), and it is predominantly ^{4}He. The ^{3}He/^{4}He ratio in the atmosphere is approximately 1.4×10^{-6}, while in natural gas, this ratio is an order of magnitude lower. Section 1.4 of this book provides a dedicated discussion of ^{3}He. Unless otherwise specified, helium mentioned hereafter refers to ^{4}He or a helium mixture where ^{4}He is the dominant isotope.

1.1.2 Properties of Helium

Helium is a monatomic gas, with a chemical symbol of He and an atomic number of 2. Its nucleus contains two protons and two neutrons, and it has two electrons orbiting the nucleus, with an atomic mass of 4.0026.

At room temperature and standard atmospheric pressure, helium is a colorless, odorless, non-toxic, non-flammable inert gas with extremely stable chemical properties. It has a normal boiling point of – 268.93 °C (4.22 K), a density of 0.1785 kg/m^3 at 0 °C, a critical temperature of – 267.96 °C (5.19 K), and a critical pressure of 0.228 MPa. Helium does not have a triple point and is the only substance that cannot solidify under standard atmospheric pressure. The basic properties of ^{4}He are shown in Table 1.1.

Liquid helium is a substance formed by liquefying helium. It has the following properties: colorless and transparent, odorless, and tasteless, making it a very pure liquid. It can reach ultra-low temperatures in the millikelvin range, serving as an important low-temperature source for scientific research. Liquid helium exhibits the λ-phase transition phenomenon, where specific heat, density, and other properties change abruptly at the λ-point. It possesses superfluidity, the creeping film

Table 1.1 Basic properties of **^{4}He** [4, 5]

Parameter	Value	Parameter	Value
Atomic mass	4.0026	Magnetic moment (nuclear spin)	0.000
Normal boiling point/K[a]	4.22	Surface tension at 0 K, N/m	3.53×10^{-4}
Density at room temperature (25 °C)/(kg/m^3)	0.1635	Specific heat capacity at 1.0 K/ 101.325 kPa, J/(kg·K)	102.3
Critical temperature Tc/K	5.1953	Vapor pressure at 1.7 K, Pa	1128
Critical pressure p_c/MPa	0.2283	Thermal conductivity of saturated liquid at 4.22 K, mW/(m·K)	18.619
Critical density ρ_c/(kg/m^3)	69.58	Thermal conductivity of saturated vapor at 4.22 K, mW/(m·K)	9.050
Minimum solidification pressure/MPa	2.5	Viscosity of saturated liquid at 4.22 K, μPa·s	3.1555
Superfluid phase transition temperature (lower λ-point[b])/K	2.178	Viscosity of saturated vapor at 4.22 K, μPa·s	1.2465
Superfluid phase transition temperature (upper λ-point)/K	1.763	Latent heat of vaporization of saturated liquid at 1 standard atmospheric pressure, kJ/kg	20.564

[a]K = C + 273.15

[b]The λ-point (Lambda point) is the temperature at which normal helium (He I) transitions to superfluid helium (He II). The value of this point varies slightly with pressure (see Sect. 1.1.3 for details)

phenomenon, and super thermal conductivity, with a viscosity coefficient close to zero.

At low temperatures, helium exists in two distinct liquid phases: the normal helium phase (He I) and the superfluid helium phase (He II). Normal helium (He I) is the usual form of liquid helium and behaves as a typical cryogenic fluid, with properties akin to liquid nitrogen, among others. Due to quantum mechanical properties, helium's nuclear spin is an even number, making it a boson that follows Bose–Einstein condensate (BEC) statistical rules. When the temperature is lowered to near absolute zero, helium undergoes BEC to transition into superfluid helium, a quantum fluid. The characteristics of superfluid helium are detailed in Sect. 1.1.3.

Helium has two electrons orbiting its nucleus, occupying the 1s orbital. These electrons have opposite spins, denoted chemically as $1s^2$. The outer electron shell of helium is completely filled, requiring no additional electrons, resulting in its complete inertness. Consequently, helium is generally considered unable to form stable compounds or be chemically bound by other substances.

In February 2017, *Nature Chemistry* published research findings on helium-sodium compounds (Na_2He) by a team from Nankai University in China, led by Zhou Xiangfeng and Wang Huitian, in collaboration with Artem Romanevich Oganov's team from Stony Brook University, USA. They successfully synthesized a thermodynamically stable helium-sodium compound, stable at pressures exceeding 113 GPa (approximately 1.1 million times Earth's atmospheric pressure). This discovery marked the end of helium's reputation as a non-compound element and provided new insights into the origins of the universe [6].

Column 1.1: Common Mass-Volume Conversions for Helium

1 m^3 of liquid helium is equivalent to approximately 700.06 m^3 of helium at 0 °C and 101.325 kPa, and about 750.96 m^3 at 20 °C and 101.325 kPa;

1 ton of liquid helium is equivalent to approximately 5602.24 m^3 of helium at 0 °C and 101.325 kPa, and about 6009.62 m^3 at 20 °C and 101.325 kPa;

1×10^4 m^3 of helium at 0 °C and 101.325 kPa is equivalent to about 1.785 tons of helium; at 20 °C and 101.325 kPa, it is equivalent to about 1.664 tons of helium.

1.1.3 Superfluid Helium

1. Properties of Superfluid Helium (He II)

Liquid helium is one of the most extensively studied fluids, with research spanning condensed matter physics, cryogenic engineering, and other fields. Superfluid helium (He II) is a special state of liquid helium that exhibits superfluid properties.

Normal helium (He I) undergoes phases such as liquefaction and solidification at low temperatures, yet at extremely low temperatures, particularly approaching absolute zero (– 273.15 °C or 0 K), liquid helium transitions into a superfluid, displaying a range of remarkable properties. In January 1938, the international journal *Nature* published two significant papers on superfluid helium (He II): *Viscosity of Liquid Helium below the λ-Point* by Soviet physicist Pyotr Leonidovich Kapitsa, and *Flow of Liquid Helium II* by Canadian scientists John Frank Allen and Don Misener. Both studies reported experimental phenomena where superfluid helium (He II) flowed without viscosity through fine channels [5].

The liquid phases of normal helium (He I) and superfluid helium (He II) are separated on a phase diagram by a phase boundary known as the λ-line, as shown in Fig. 1.2. Normal helium (He I) is a typical viscous fluid, and the transition to superfluid helium (He II) occurs at the λ-point, $T_{\lambda} = 2.178$ K, under saturated pressure. This point is referred to as the lower λ-point, and He I exists as a liquid phase above the λ-point temperature. Superfluid helium (He II) is a superfluid exhibiting quantum mechanical characteristics and can be theoretically explained using BEC theory.

Early experiments demonstrated that superfluid helium (He II) has extremely low viscosity. Compared to normal helium (He I), its viscosity coefficient is several orders of magnitude lower. Due to its low viscosity, experiments require stringent containment measures, as any leaks would result in the rapid disappearance of superfluid helium (He II), hindering experimental proceedings.

Fig. 1.2 Phase diagram of ^{4}He

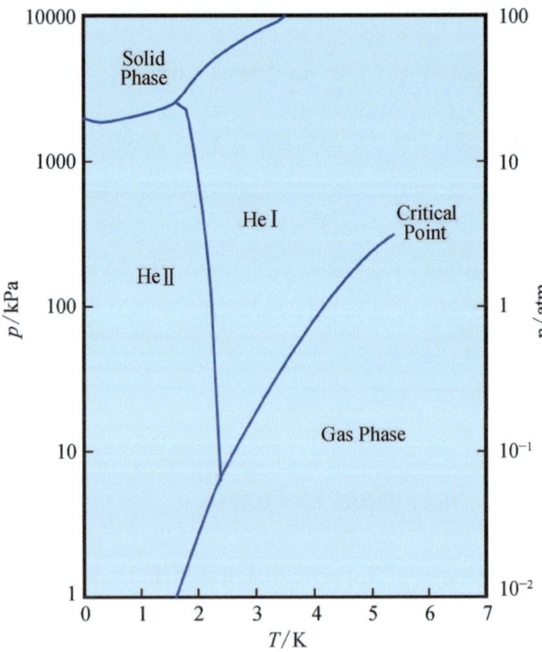

2. Peculiar Phenomena of Superfluid Helium (He II)

Superfluid helium (He II) exhibits the intriguing creeping film phenomenon. This phenomenon involves a 30 nm-thick film of liquid helium in the He II state, known as the Rollin film, named after Bernard V. Rollin, as illustrated in Fig. 1.3. A small container is partially immersed in a larger container, both filled with superfluid helium (He II). Initially, the small container has a lower liquid level than the larger container, yet as the experiment progresses, it gradually fills with superfluid helium (He II) until the levels equalize. When the liquid level of the small container is raised above that of the larger container, the superfluid helium (He II) inside it decreases until the levels match. If the liquid level in the small container is elevated to a point where it is completely above that in the larger container, the superfluid helium (He II) inside the small container gradually decreases until it is completely empty.

The creeping film phenomenon can be explained by the weak interaction between helium atoms and their strong attraction to glass surfaces, causing atoms to continuously seek new surfaces to adhere to. Moreover, the extremely low viscosity of superfluid helium (He II) makes further creeping possible.

Another peculiar phenomenon of superfluid helium (He II) is the thermomechanical effect, as depicted in Fig. 1.4. In experiments, a porous medium is placed inside a glass tube with pore diameters < 0.1 μm, and the tube's ends are sealed with cotton. When a light beam (e.g., from a flashlight) illuminates the porous medium, superfluid helium (He II) is observed to spout from the top of the glass tube like a fountain, hence this effect is also called the fountain effect. The experimental setup includes a heater installed at the porous medium, and when heating superfluid helium (He

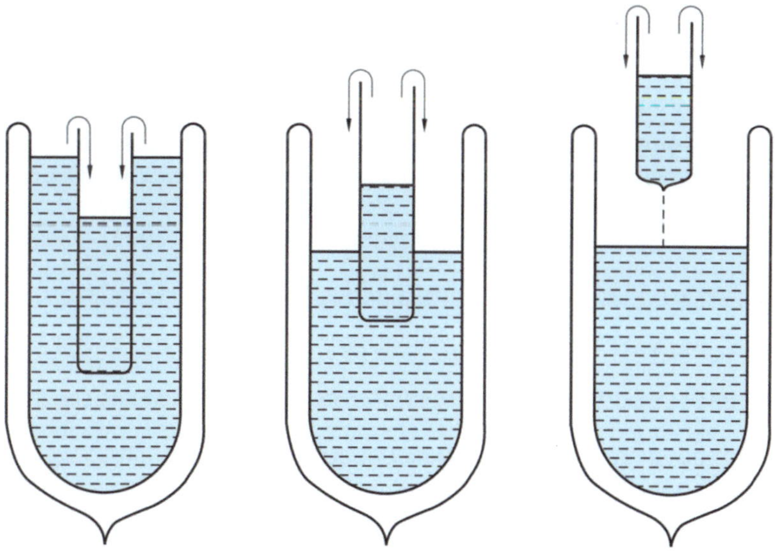

Fig. 1.3 Creeping film phenomenon in superfluid helium (He II)

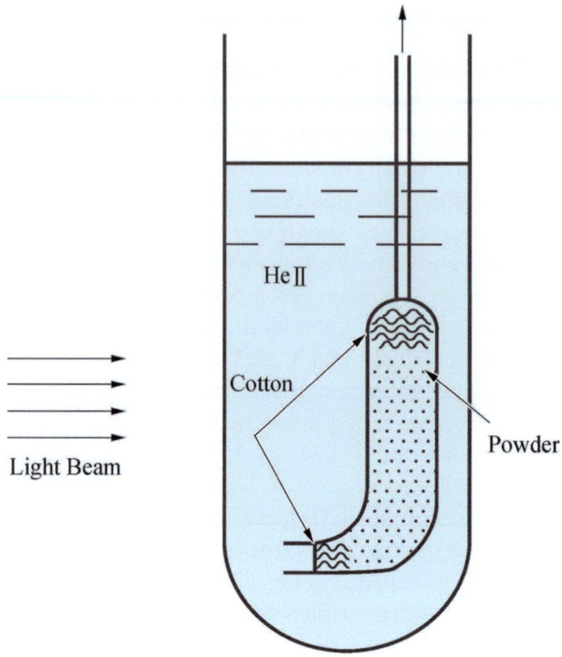

Fig. 1.4 Schematic diagram of thermomechanical effect in superfluid helium (He II)

II), the thermomechanical effect arises. If the setup and operation are error-free, the fountain can reach heights of several tens of centimeters, as shown in Fig. 1.5.

3. Applications of Superfluid Helium (He II)

Due to its low viscosity and high thermal conductivity, among other properties, superfluid helium (He II) is an excellent coolant with superior cooling performance, widely used in various fields. In space exploration, for instance, superfluid helium (He II) is often used to maintain the low-temperature environment required for detectors, so as to boost the sensitivity of space-based far-infrared detectors. In 2003, the NASA Infrared Telescope Facility (NASA IRTF) used approximately 400 L of superfluid helium (He II). Superfluid helium (He II) is also commonly used to cool superconducting magnets to improve thermal stability, achieve strong magnetic fields, and secure system operation. The Large Hadron Collider (LHC) of the European Organization for Nuclear Research (CERN), for example, uses large quantities of superfluid helium (He II) to cool its superconducting magnets. Additionally, in condensed matter physics research, the unique properties of superfluid helium (He II) are a focal point for low-temperature physicists, involving studies on quantum vortices, solid-state superfluidity, and so forth [8, 9].

Fig. 1.5 Fountain effect in superfluid helium (He II) [7]

1.1.4 Origin of Helium

On Earth, helium mainly exists as two stable isotopes: ^3He and ^4He. It is generally believed that ^3He is a primordial helium isotope that has been present in the Earth's interior since its formation, whereas ^4He is produced through radioactive processes. Over Earth's long geological history, ^3He and ^4He have diffused to the surface through fault zones. However, due to radioactive decay, ^4He has gradually accumulated in the Earth's crust. The intensity of fault activity, the permeability coefficient of the medium, and the sealing conditions often vary in different geological structures, leading to significant differences in the $^3He/^4He$ and $^4He/^{20}Ne$ values in geological samples.

Chris J. Ballentine and others have proposed that there are three main sources of helium on Earth: atmospheric, crustal (radiogenic), and mantle sources [10]. Helium of atmospheric sources mainly enters the basin fluid system through the circulation of groundwater. In groundwater recharge zones, helium from the atmosphere dissolves into the recharge water and migrates from the recharge zone to the discharge zone. It then reaches oil and gas reservoirs, where it is exchanged among oil, gas, and water, leading to helium accumulation in these reservoirs. Helium from crustal sources is produced by the radioactive decay of elements like uranium (U) and thorium (Th) in minerals and rocks, primarily resulting in ^4He. Helium of mantle sources is released from the mantle through degassing processes associated with magma activities, primarily contributing ^3He.

1. Atmospheric Helium

Helium in the atmosphere is primarily released by volcanic eruptions along mid-ocean ridges, magma degassing, and rock weathering. Helium concentration in the atmosphere is extremely low, about 0.000524%, and so is the amount of helium dissolved in water in recharge areas and then entering basin fluid systems through groundwater circulation—which is often negligible. Therefore, helium in subsurface fluids mainly originates from crustal and mantle sources.

2. Crustal ^4He

The radioactive decay (α decay) of uranium-235 (^{235}U), uranium-238 (^{238}U), and thorium-232 (^{232}Th) produces ^4He and β particles. The primary radioactive decay reactions are as follows:

$$^{238}_{92}U \rightarrow 8^4_2He + 6\beta + ^{206}_{82}Pb \tag{1.1}$$

$$^{235}_{92}U \rightarrow 7^4_2He + 4\beta + ^{207}_{82}Pb \tag{1.2}$$

$$^{232}_{90}Th \rightarrow 6^4_2He + 4\beta + ^{208}_{82}Pb \tag{1.3}$$

The yield of crustal helium depends on the content of U and Th in rocks. Higher content result in more helium output from radioactive decay. Almost all rocks in the Earth's crust contain U and Th, but their concentrations are low, and their half-lives are exceedingly long: ^{235}U has a half-life of 704.8 million years, ^{238}U has 4.468 billion years, and ^{232}Th has 14.01 billion years [11]. As a result, helium is produced at an extremely slow rate. Part of the generated helium escapes into the atmosphere, while the remainder diffuses within the lithosphere, making it challenging to accumulate into independent helium reservoirs. To date, no independent crustal helium reservoirs have yet been discovered worldwide.

3. Mantle ^3He

Significant amounts of mantle-derived ^3He can be found in basins and oil and gas reservoirs. The presence of ^3He in the mantle suggests tectonic activity at the base of the basin, as only mantle melting and magma activity can continuously transport ^3He and other volatiles into the basin fluid system.

The $^3He/^4He$ values vary significantly among different sources, which allows for the determination of helium's origin. Typically, the $^3He/^4He$ value for atmospheric sources is around 1.4×10^{-6} [12], while for crustal sources, the value is approximately 2×10^{-8} [13]. In contrast, the value for mantle sources is about 1.1×10^{-5} [12].

By measuring the $^3He/^4He$ values, researchers can study the distribution of helium isotopes in different geological structures, revealing differences in the intensity of tectonic activities. This measurement can also be used to trace the characteristics of helium isotopes from different material sources, such as rocks, groundwater, and the atmosphere, helping to determine the origin and evolution of samples. Additionally, studying the migration processes of ^3He and ^4He provides insight into the mechanisms of gas transport within the Earth's interior.

4. Helium in Natural Gas

Helium is often associated with natural gas (shale gas, coalbed methane, nitrogen) or geothermal water, and independent helium reservoirs have not yet been discovered worldwide. Compared to hydrocarbon natural gases, helium is a typical weak-source gas. Helium-source rocks exhibit low gas generation intensity, making it challenging for large-scale, concentrated gas discharges akin to hydrocarbon natural gases that breach the "pressure compartment". This difficulty in forming free gas columns driven by buoyancy to enter traps and create helium reservoirs renders helium less likely to accumulate independently. Effective helium-source rocks, efficient migration pathways, and carrier gas reservoirs are essential for helium accumulation. After formation in helium-source rocks, helium mixes with other carrier gases or dissolves in groundwater. Through efficient transport pathways such as faults, helium is then conveyed to natural gas reservoirs and other carrier gas deposits over long distances on a large scale, creating helium-rich reservoirs [14].

1.1.5 Research History of Helium

Helium, due to its unique properties, has become one of the most studied elements by scientists. Since the twentieth century, research on helium has provided humanity with fundamental new knowledge and achievements, metaphorically described as the conquest of a series of "peaks" [15, 16]:

(1) In 1908, the team led by physicist Heike Kamerlingh Onnes at Leiden University in the Netherlands liquefied helium for the first time, with a temperature of 4.22 K (– 268.93 °C).
(2) In 1911, the same team discovered superconductivity in mercury at liquid helium temperatures, marking the dawn of cryogenic engineering and low-temperature physics.
(3) In 1926, Willem Hendrik Keesom solidified helium for the first time.
(4) In 1934, Marcus Laurence Elwin Oliphant proposed the existence of ^3He.

(5) In 1934, Pyotr Leonidovich Kapitsa liquefied helium using the gas expansion cycle (piston expander).
(6) In 1938, Pyotr Leonidovich Kapitsa discovered superfluidity in ^{4}He, known as superfluid helium (He II).
(7) In 1938, Fritz Wolfgang London proposed BEC as the mechanism for superfluidity and superconductivity in ^{4}He.
(8) In 1939, Luis Alvarez and Robert Cornog successfully isolated ^{3}He for the first time [17].
(9) In 1949, Darrell W. Osborne, Bernard Weinstock, and Bernard M. Abraham, among others, conducted pioneering research on liquid ^{3}He [5], discovering that liquid ^{3}He is not a superfluid at temperatures near T_λ [9].
(10) In 1950, Vitaly Lazarevich Ginzburg and Lev Davidovich Landau proposed the Ginzburg–Landau Theory, which describes superconductivity.
(11) In 1951, Darrell W. Osborne and others solidified ^{3}He [18].
(12) In 1955, George Yntema successfully constructed the first superconducting magnet, a small 0.7 T iron-core electromagnet with a superconducting niobium winding, ushering in the era of engineering superconductivity.
(13) In 1957, John Bardeen, Leon N. Cooper, and John Robert Schrieffer developed the BCS theory (Bardeen-Cooper-Schrieffer theory), which describes superconductivity as a microscopic effect resulting from Cooper pair condensation. This was the first microscopic theory of superconductivity since Heike Kamerlingh Onnes' great discovery in 1911.
(14) In 1962, Heinz London and Kurt Mendelssohn, among others, proposed practical techniques for dilution refrigeration.
(15) In 2017, a team led by Wang Huitian and Zhou Xiangfeng from Nankai University, along with collaborators, published a paper in *Nature Chemistry* about the synthesis of sodium-helium compound (Na_2He) under high-pressure conditions, ending the long-standing belief that helium cannot form compounds [6].

Research on helium has not only expanded our understanding of the universe but also propelled scientific and technological advancements. Researchers in the field of helium have also been duly recognized with numerous honors. Table 1.2 lists 16 Nobel Prize laureates who have made significant contributions to helium research over a span of more than 90 years, illustrating that helium has been a hot topic of research for over a century, with new phenomena, theories, and achievements continuously emerging.

1.2 Typical Uses of Helium

Since its discovery over 100 years ago, helium has been increasingly used in a variety of fields due to its unique physical and chemical properties. In the first half of the twentieth century, one of the significant uses of helium was as a substitute for

Table 1.2 Nobel laureates and their contributions to helium research

No.	Year	Name	Award	Contribution
1	1904	William Ramsay, UK Chemist	Nobel Prize in Chemistry	Discovery of the inert gaseous elements in air, and his determination of their place in the periodic system
2	1913	Heike Kamerlingh Onnes, Netherlands Physicist	Nobel Prize in Physics	Investigations on the properties of matter at low temperatures which led, inter alia, to the production of liquid helium
3	1962	Lev Davidovich Landau, USSR Physicist	Nobel Prize in Physics	Pioneering theories for condensed matter, especially liquid helium
4	1972	John Bardeen, USA Physicist; Leon N. Cooper, USA Physicist; John Robert Schrieffer, USA Physicist	Nobel Prize in Physics	Jointly developed theory of superconductivity, usually called the BCS-theory
5	1978	Pyotr Leonidovich Kapitsa, USSR Physicist	Nobel Prize in Physics	Basic inventions and discoveries in the area of low-temperature physics (superfluidity in ^4He)
6	1996	David Morris Lee, USA Physicist; Douglas Dean Osheroff, USA Physicist; Robert Coleman Richardson, USA Physicist	Nobel Prize in Physics	Discovery of superfluidity in ^3He
7	2001	Eric Allin Cornell, USA Physicist; Carl Edwin Wieman, USA Physicist; Wolfgang Ketterle, Germany Physicist	Nobel Prize in Physics	Achievement of BEC in dilute gases of alkali atoms, and for early fundamental studies of the properties of the condensates
8	2003	Alexei Alexeyevich Abrikosov, Russia & USA Physicist; Vitaly Lazarevich Ginzburg, Russia Physicist; Anthony James Leggett, UK & USA Physicist	Nobel Prize in Physics	Pioneering contributions to the theories of superconductors and superfluids

Table 1.3 Properties and applications of helium

No.	Property	Application
1	Extremely low boiling point, does not solidify at 0 K under standard atmospheric pressure	Cryogenic superconductivity, purging liquid hydrogen systems
2	Second lightest element after hydrogen	Lifting medium for balloons and airships
3	Very small molecular size, highest permeability	Leak detection, supersonic wind tunnels (high Reynolds number)
4	Chemical inertness	Carrier gas, analytical gas, semiconductors
5	Very high specific heat and thermal conductivity	Rapid cooling in vacuum furnaces, gas fiber optic cooling, nuclear reactors
6	Radioactive inertness	Heat transfer medium in nuclear fission/fusion reactors
7	Highest ionization potential	Metal arc and plasma arc welding
8	Extremely low solubility in water	Deep-sea diving gas
9	Very high speed of sound	Metal coating

flammable and explosive hydrogen to provide buoyancy for balloons and airships, contributing to airships becoming the first transoceanic manned transport. Since 1949, when helium with a purity greater than 99.0% began to be produced, high-purity helium products have expanded the scope of its applications, particularly in fields such as cryogenic superconducting, aerospace, optical fibers and semiconductors, welding protection, chromatography, leak detection, and breathing gas mixtures. Table 1.3 lists some properties of helium and its main applications.

Column 1-2: The Hindenburg Disaster

On May 6, 1937, the Hindenburg airship of Zeppelin Corporation burned up in just 34 s while preparing to land at the Lakehurst Naval Air Station in New Jersey. The disaster occurred at an altitude of about 245 m, witnessed by over a thousand spectators who saw the Hindenburg turn into a skeleton of flames for an instant. Due to the U.S. restrictions on helium exports, the Hindenburg used hydrogen instead of helium for buoyancy. During the landing maneuver, cables tore the airship's number 4 gas cell, causing a hydrogen leak. Prior to landing, a thunderstorm had just passed, accumulating a significant amount of static electricity on the airship, which ignited the leaking hydrogen, resulting in a massive fire. All 2.15×10^5 m^3 of hydrogen in the 16 gas cells burst into flames.

The Hindenburg was the largest aircraft ever built, weighing about 110 tons and measuring 245 m in length, 3.5 times the size of a Boeing 747 and only 24 m shorter than the Titanic. Before the disaster, it had completed 63 flights, including 37 transatlantic crossings. It was capable of flying 14,000 km at a speed of 290 km/h. During this transatlantic flight against headwinds, the Hindenburg was delayed by twelve hours. As it slowly descended over the East Coast, it sought to land in the evening. Around 7 p.m., the airship dropped two landing ropes in preparation for landing. Then, at an altitude of 245 m, the massive airship suddenly caught fire. People on the ground were at a loss, helplessly watching the hydrogen cells almost instantly consumed by flames. The burning framework crashed to the ground and was shattered, with the Hindenburg consumed by smoke. Through the efforts of naval personnel and crew members, many people were rescued. However, 36 of the 97 passengers and crew members perished in the disaster.

The Hindenburg disaster marked the end of the commercial airship era. The air transport industry changed its course following the blow from the Hindenburg. In 1939, the first commercial airplane carrying passengers crossed the Atlantic, and airplanes replaced airships as the main means of air transportation.

As a matter of fact, Zeppelin Corporation, which owned the Hindenburg, had long been aware of the risks of using hydrogen. When designing the Hindenburg, the original plan was to replace hydrogen with helium. During World War I, the U.S. government realized the importance of helium for aircraft and issued the *Helium Act* in 1925 as a result, establishing a national helium reserve in Amarillo, Texas, and prohibiting the export of helium. Under the influence of this act, Zeppelin Corporation struggled to obtain helium and, out of necessity, had to take the enormous risk of using hydrogen as the filling gas for the Hindenburg, a decision that foreshadowed this disaster.

Hindenburg disaster [19]

Column 1-3 Liquid Helium Aiding Nuclear Magnetic Resonance

Nuclear magnetic resonance (NMR) is a physical phenomenon widely used as an analytical method in physics, chemistry, and biology, and since 1973, it has been used in medical clinical diagnostics.

During an MRI scan, the patient is moved into a large cylindrical magnet that is extremely powerful. Under the influence of radiofrequency pulses at different frequencies, your body is exposed to different magnetic fields. Different atoms absorb the magnetic field's energy differently, and some atoms are more likely to absorb the energy than others, allowing for clear cross-sectional images that enable doctors to see tumors and other abnormalities. Therefore, NMR serves as a magical key, helping doctors better understand your body's condition.

In this process, liquid helium, as a refrigerant, plays a significant role by maintaining the superconducting state of the magnets, allowing for higher sensitivity and resolution in NMR measurements while effectively reducing operating costs. In the structure of an NMR machine, the superconducting coils and magnet housing are completely enclosed within a container filled with liquid helium, maintaining a temperature of 4 K, which gives the coils their superconducting properties.

In a 3.0 T superconducting MRI cooling system, for instance, the superconducting coils generate a magnetic field when powered, maintaining an ultra-low temperature environment where the magnetic field can persist for an extended period, thereby achieving the desired imaging results.

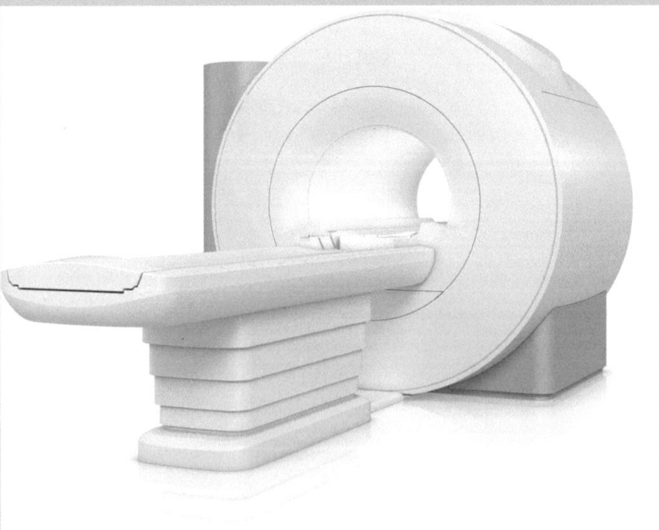

Magnetom Cima. X NMR

1.2.1 Cryogenic Superconductivity

In the field of medical imaging, Nuclear Magnetic Resonance Imaging (NMRI) relies heavily on liquid helium as a coolant. This allows superconducting materials within the equipment to be cooled to superconducting states, achieving higher sensitivity and resolution in measurements. Medical NMRI represents one of the largest consumers of liquid helium. Additionally, liquid helium refrigerators are utilized to achieve and maintain low-temperature conditions in major scientific installations, such as electron–positron colliders, synchrotron radiation light sources, spallation neutron sources, and astronomical telescopes. The demand for helium in the cryogenic superconductivity field currently accounts for about 40% of total helium consumption.

1. Nuclear Magnetic Resonance

Liquid helium enables the superconducting magnets inside NMRI scanners to remain at ultra-low temperatures, therefore producing very strong magnetic fields. NMR machines vary in size and require between 500 and 1000 L of liquid helium each time they are filled, with refills necessary 1–2 times per year. In recent years, the global NMR equipment market has experienced stable growth due to increasing research and clinical needs. In China, the number of NMR units has risen from 4376 in 2013 to 13,242 in 2022, leading to significant demand for liquid helium. The demand for liquid helium in the medical field is expected to continue growing rapidly in the future.

2. Electron–Positron Collider

The Beijing Electron–Positron Collider (BEPC) is a significant particle physics experimental facility established by the Institute of High Energy Physics, Chinese Academy of Sciences (IHEP of the CAS), in October 1988. It is China's first high-energy accelerator and is considered one of the world's eight major high-energy accelerator centers, as shown in Fig. 1.6. BEPC is primarily used for electron–positron collision experiments, allowing the study of the fundamental structure and properties of particles through high-energy collisions. Since its operation began in 1990, BEPC has achieved numerous breakthroughs in precision measurements, such as the precise measurement of hadron reaction cross-sections (R-values) in the 2 billion to 5 billion electron volt energy range, significantly impacting the international high-energy physics community. BEPC also serves as a large public experimental platform for synchrotron radiation in China, facilitating interdisciplinary research in fields like condensed matter physics, materials science, biology and medicine, environmental science, geological and mineral resources, and microfabrication technology.

In the upgraded version, BEPC-II, the superconducting system is widely applied. First, as superconducting magnets, they create high-strength magnetic fields near the collision points, focusing and guiding the electron and positron beams to ensure high-energy collisions. Second, they are used in the RF cavities of the electron accelerators. Superconducting RF cavities operate at ultra-low temperatures to reduce energy

Fig. 1.6 Beijing electron–positron collider. *Source* Official Website of the IHEP of the CAS

losses and provide higher acceleration efficiency, enabling particles in the accelerator's circular orbit to reach higher speeds. The superconducting system requires ultra-low temperature operation, so BEPC-II is equipped with a helium cryogenic refrigerator providing approximately 800 W of cooling capacity at 4.5 K.

3. Synchrotron Radiation Light Source

The Synchrotron Radiation Light Source is an advanced laboratory light source that utilizes the interaction between electron beams and magnetic fields to produce high-brightness, high-intensity light radiation. It is commonly used to study the structure, properties, and functions of materials and to conduct experiments in fields such as biomedical science, materials science, and chemistry.

Synchrotron radiation light sources have undergone three generations of development, each with different performance characteristics. The first generation primarily served as dual-purpose machines for high-energy colliders, such as the Beijing Synchrotron Radiation Source (BSR), which was parasitic on the BEPC. The second generation consisted of dedicated storage rings for synchrotron radiation, such as the Hefei Light Source (HLS) at the National Synchrotron Radiation Laboratory. The third generation also features dedicated storage rings but with higher performance, such as the Shanghai Synchrotron Radiation Facility (SSRF). The development of each generation of radiation light sources represents progress in synchrotron radiation technology, with differences in electron beam spot size or electron emissivity serving as the luminescent light source.

Fig. 1.7 Shanghai synchrotron radiation facility. *Source* Official Website of the CAS

Superconducting technology is widely applied in the storage rings and RF systems of synchrotron radiation light sources. The storage ring is the core component of a synchrotron radiation light source. By using superconducting electromagnets to manufacture the storage ring's magnets, stable electromagnetic fields can be created, thereby elevating electron beam stability and beam quality. The RF system in the storage ring maintains the electron beam's energy, with superconducting RF cavities providing higher electric field strengths. This allows the electron beam to be more efficiently accelerated, increasing the synchrotron radiation light source's brightness and stability. In the case of the SSRF, as shown in Fig. 1.7, it includes three superconducting RF cavities (Superconducting Radio Frequency, SCRF, 499.65 Hz/40 MV) and utilizes a helium cryogenic refrigerator with a cooling capacity of 650 W at 4.5 K.

4. Spallation Neutron Source

Spallation Neutron Sources (SNS) are devices that produce neutron beams and are extensively used in various fields such as materials science, physics, chemistry, and biology. They provide scientific research with a powerful tool that plays a significant role in studying the microscopic structure and properties of matter.

Prominent examples of facilities using cryogenic technology for spallation neutron sources include the ISIS Neutron and Muon Source in the UK, the Spallation Neutron Source in the USA, the Japan Proton Accelerator Research Complex (J-PARC), and the China Spallation Neutron Source (CSNS). The CSNS (as shown in Fig. 1.8), located in Dongguan, Guangdong, is composed of a negative hydrogen linear accelerator, a rapid cycling proton synchrotron, a beam transport line, a target station, a neutron spectrometer, and various supporting facilities. The cryogenic system is a key component of the CSNS. It consists of two closed-loop systems: a

Fig. 1.8 A weapon of scientific research—China spallation neutron source. *Source* Official Website of the CAS

supercritical hydrogen circulation system for neutron cooling and a helium refrigeration system for cooling hydrogen. The system provides the neutron moderator with a cooling capacity of about 2000 W at around 20 K.

5. Astronomical Telescope

When astronomical telescopes are used to detect distant objects, the noise from electronic devices can interfere with the ground-based receiving system's ability to capture signals. The noise is primarily caused by the random motion of electrons. Theoretical studies have shown that the noise temperature of deep-space receiving systems is a critical factor affecting data reception rates. Reducing the noise temperature by 50% can double the data reception rate, thereby effectively increasing the deep-space detection range. At 20 K, the equivalent noise of astronomical telescopes is only about 10%–20% of what it is at room temperature.

The world's largest single-aperture radio telescope, known as the Five-hundred-meter Aperture Spherical Radio Telescope (FAST), is located in Pingtang, Guizhou, China, as depicted in Fig. 1.9. Built by the National Astronomical Observatories, CAS, FAST has a diameter of 500 m, covering an area equivalent to about 30 football fields, making it one of the world's most sensitive radio telescopes. It was completed in September 2016 and officially put into use in January 2020. The ultra-wideband receiver system is an essential component of the FAST system, consisting of an

Fig. 1.9 The world's largest single-aperture telescope—FAST. *Source* Xinhua News Agency

ultra-wideband feed, cryogenic Dewar, low-noise amplifier system, and RF receiver circuit system. The cryogenic Dewar and low-noise amplifier system include a helium refrigerator that can maintain the temperature of the Dewar's second-stage cold head below 15 K [9].

Column 1-4: Superconductivity Phenomenon

In 1908, Dutch physicist Heike Kamerlingh Onnes became the first person to liquefy helium. In 1911, he immersed materials in liquid helium and discovered that mercury entered a new state at temperatures around 4.22 K, where its resistance became immeasurably small. Surprised by this characteristic, he repeated the experiment several times, only to obtain the same result. He called this new state of mercury the superconducting state.

Conductors in the superconducting state are later referred to as superconductors, characterized by the sudden disappearance of DC electrical resistance at certain low temperatures, known as the zero-resistance effect. Without resistance, a superconductor can carry large currents without heat loss, creating extremely strong magnetic fields. In the field of superconductivity, liquid helium is extensively used to achieve superconductivity. Superconducting materials in this state exhibit zero electrical resistance and complete magnetic flux expulsion, making them important for scientific research and industrial applications.

Helium is unique in that it does not solidify even near "absolute zero", and its low boiling point makes it one of the ideal elements for achieving superconductivity. This property is based on the presence of van der Waals forces, named after Dutch scientist Johannes Diderik van der Waals, which are attractive forces between molecules (or atoms). The strength of these forces is positively correlated with the size of the molecules (or atoms). The van der Waals forces between helium atoms are very weak, especially the London dispersion force, the weakest of van der Waals forces. At extremely low temperatures, helium atoms form fluctuating dipoles that attract each other, ultimately leading to liquefaction.

1.2.2 Aerospace

Helium plays an extremely important role in the aerospace industry. It is used for pressurizing rocket propellants, as a propellant or booster gas for satellite and other spacecraft attitude and orbit control systems, and for leak detection in various pipelines, valves, and pressure vessels on rockets and spacecraft.

Given its extremely low boiling point, low density, non-reactivity with propellants, and insolubility in propellants, helium is widely used for pressurizing rocket propellants. China's Long March 3, Long March 4, Long March 5, and Long March 7 series of launch vehicles have adopted different types of helium pressurization technologies as needed. Helium pressurization systems are also used in the United States' Saturn V and Atlas V, the Soviet Union's Zenit, Europe's Ariane 5, Russia's Angara, and Japan's H-II rockets. Before a cryogenic-fueled rocket engine is ignited, it must undergo pre-cooling to acclimate the entire engine to the cryogenic liquid fuel. This prevents serious problems that can severely affect rocket engine operation, such as pipeline gas trapping and cavitation in critical components. Thus, helium is used not only for propellant pressurization but also for engine precooling and blowing. Space X's Falcon 9 rocket makes extensive use of helium. Its first stage is equipped with eight helium cylinders, which are successively used for tank pressurization, cold gas attitude control, engine start blowing and precooling, cold gas separation, and other functions.

Figure 1.10 illustrates the Long March 7 launch vehicle. Its workflow is as follows: (1) The cargo bay carries satellites and other supplies into orbit. (2) The second stage is equipped with liquid oxygen and kerosene. (3) Four engines consume 71.7 tons of propellant. (4) The boosters separate via pyrotechnic separation. (5) The total liftoff thrust reaches 7080 kN.

Fig. 1.10 Long March 7 rocket

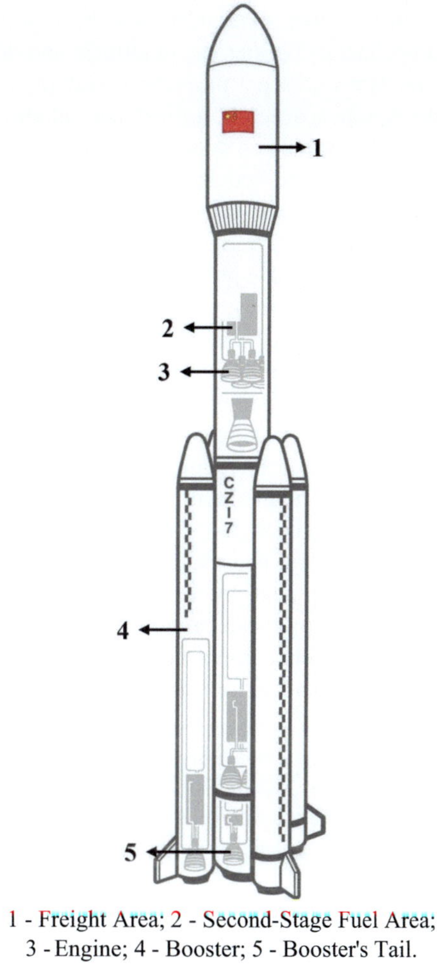

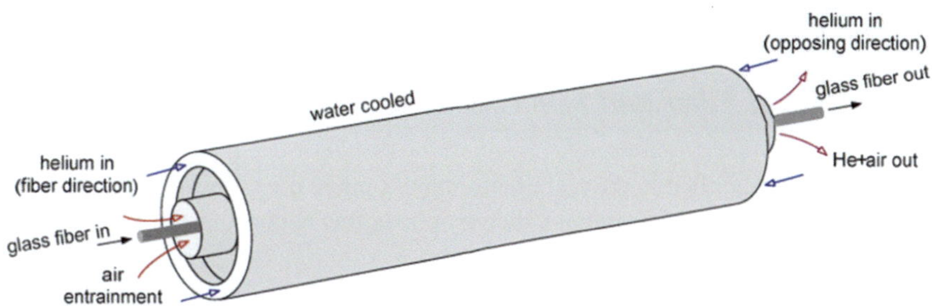

Fig. 1.11 Glass fiber cooling apparatus supplying helium

In satellites, spacecraft, and space probes, helium is also commonly used as a propellant or booster gas in attitude and orbit control systems. The descent propulsion system, ascent propulsion system, and service module propulsion system of the Apollo spacecraft, for instance, all utilized helium pressurization systems. Similarly, the propulsion systems of China's Chang'e 5 lunar probe and Tianwen-1 Mars probe employed helium pressurization technology. Helium is also used to provide the extremely low temperatures required for the operation of key components in spacecraft. In 2009, the European Space Agency (ESA) launched a spacecraft called Planck, designed to probe the cosmic microwave background radiation. After a series of cooling processes, a helium dilution refrigerator cooled its detectors to 0.1 K. This was the first time humanity used a fully active cooling system without liquid helium precooling to achieve millikelvin temperatures in space. The Alpha Magnetic Spectrometer, developed under the leadership of Nobel laureate and Chinese-American physicist Tsung-Dao Lee to detect cosmic antimatter, also utilized superfluid helium technology for its cooling apparatus.

In space environment simulation facilities, an important device that simulates the cold dark background of the universe is called "thermal sink". Spacecrafts must undergo testing and evaluation in space environment simulation facilities during their development process. To meet the requirements for high discharge manned spaceflight and rocket engine plume tests, liquid helium thermal sinks with stronger cooling capacity must be used to keep the entire thermal sink temperature below 10 K. Additionally, to address the need for extremely high and low temperatures and large temperature variations required by lunar or deep space exploration spacecraft, the "helium-atmosphere space environment simulation technology" has been developed for use in test facilities.

The pipelines, valves, and pressure vessels on rockets and spacecraft play a pivotal role in space missions. Leaks in these components have led to numerous major space accidents. Therefore, testing for leaks in pipelines, valves, and pressure vessels is a critical step in the manufacturing process of rockets and spacecraft. Currently, leak detection technology using helium mass spectrometers is widely applied in the aerospace industry, where it assumes a significant role.

1.2.3 Optical Fiber and Semiconductor

Helium is widely applied in the manufacturing of optical fibers and semiconductors. It is used to control the environment, thereby preventing exposure to ambient air that could diminish product performance.

In optical fiber manufacturing, helium's high thermal conductivity and inertness are of great importance to cooling glass fibers drawn from the furnace, as shown in Fig. 1.11. Helium is also essential during the coating process of the glass fibers. The first step in optical fiber production is creating glass preforms, where hydrogen, oxygen, and helium are used to build up the strength of the glass rods for heating and stretching later. Helium and argon help rapidly cool the heated and stretched

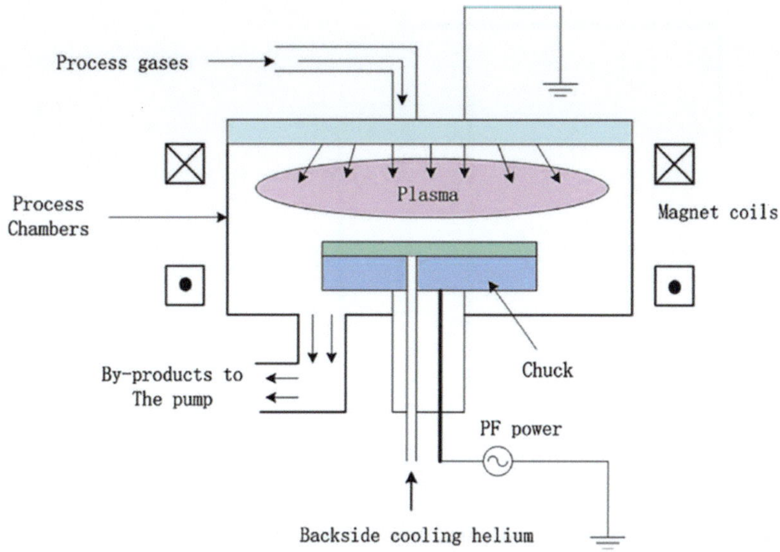

Fig. 1.12 Schematic diagram of plasma etching chamber [20]

optical fibers. Without helium, bubbles might come into being, rendering the fibers useless. Currently, with the development of 5G networks, the demand for optical fibers in China is expected to grow continuously, leading to a steady increase in helium demand in the optical fiber industry.

In semiconductor manufacturing, helium's chemical inertness and high thermal conductivity make it an ideal protective atmosphere for silicon wafers. It helps reduce wafer temperature and improve yield. Additionally, helium serves as a carrier gas in deposition processes, such as diluent gas in plasma etching and specific chip cooling applications. It is also applied to ultra-clean manufacturing and assembly environments, which are critical for semiconductor manufacturing (Fig. 1.12).

1.2.4 Welding Protection

Helium is an inert gas commonly used as a shielding gas in welding applications to prevent contamination of the weld area by atmospheric oxygen and water vapor. This helps to avoid issues such as metal oxidation and hydrogen-induced cracking, thereby improving the quality of the weld. Additionally, introducing helium as a shielding gas effectively reduces spatter during the welding process, lowers the porosity rate, and helps refine the molten droplets in the weld area, securing a more uniform distribution across the area. By adjusting the flow rate and mixture of helium, the depth of weld penetration can be more precisely controlled to ensure uniformity in the welded area.

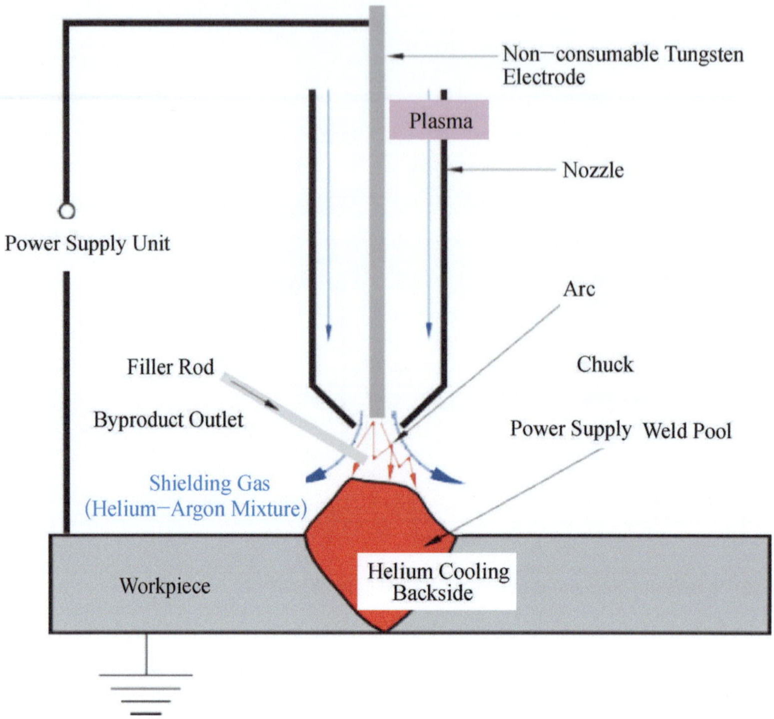

Fig. 1.13 Schematic diagram of TIG [21]

In addition, due to its excellent thermal conductivity, helium enables the weld area to reach high temperatures more quickly, which aids in the melting and penetration of the weld and increasing the welding speed. For example, in Tungsten Inert Gas (TIG) welding, the heat generated by the arc between the tungsten electrode and the workpiece melts the base material and filler wire (if used) under the protection of an inert gas. During welding, the shielding gas is continuously emitted from the nozzle of the welding gun, creating a gas shield around the arc to isolate it from the atmosphere. This prevents any detrimental effects on the tungsten electrode, weld pool, and adjacent heat-affected zone, thereby yielding high-quality welds. The shielding gas can be argon, helium, or a mixture of argon and helium. The TIG is shown in Fig. 1.13. The proportion of helium-argon mixture can be adjusted according to specific welding requirements.

1.2.5 Chromatographic Analysis

Chromatographic analysis is widely used across various industries such as pharmaceuticals, food, and environmental analysis to test and analyze the composition

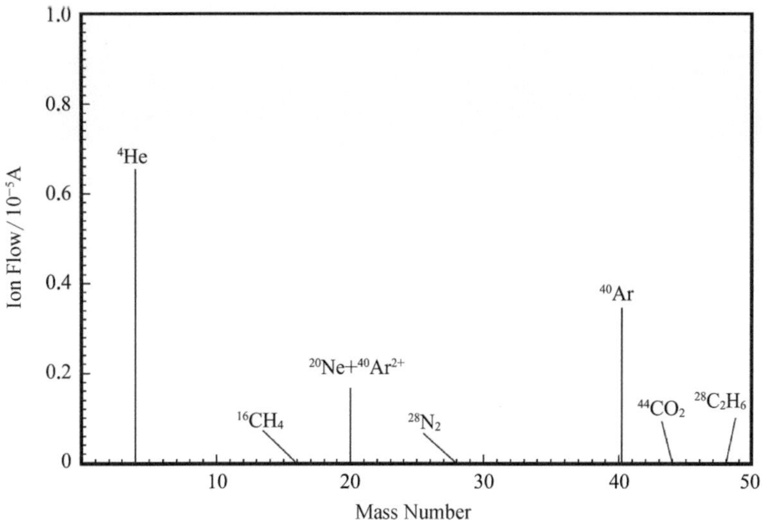

Fig. 1.14 Chromatographic analysis diagram

of samples. Chromatography is one of the methods to analyze trace elements and compounds in complex mixtures. In chromatographic applications, the column is a critical part of the chromatographic separation process. Helium is often used as a carrier gas to propel compounds through the chromatographic column, taking advantage of the differing adsorption forces of the adsorbent within the chromatographic column to separate the components of the mixture. Due to helium's inertness and low molecular weight, it does not chemically react with the compounds in the sample, thereby preserving the purity and securing the separation. Moreover, with excellent diffusion properties, helium is able to spread evenly, which helps improve the efficiency and accuracy of chromatographic separation. Its low viscosity minimizes pressure drop through the adsorbent, contributing to better separation and prolonged chromatographic column life. For instance, gas chromatography (GC) can be used to test the composition of natural gas and thus to determine its constituents for processing and usage purposes (Figs. 1.14 and 1.15).

1.2.6 Leak Detection Gas

Helium is an ideal element for testing small leaks or microcracks due to its smallest atomic radius and complete inertness. It is commonly used as a leak detection gas in high-end equipment. Helium leak detection is widely applied in various commercial applications, such as locating small leaks in automotive and aircraft fuel tanks, fuel systems, engines, refrigerator compressors, and light bulbs, among others. These

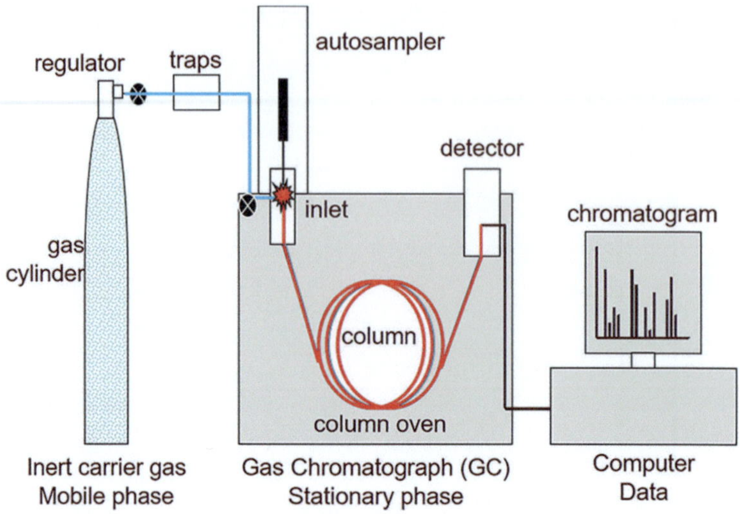

Fig. 1.15 Simplified schematic diagram of gas chromatography [22]

applications help prevent adverse impacts on public safety. Helium is also used to identify potential leaks in natural gas or petroleum pipelines.

The method for using helium to detect product leaks involves placing the product in a vacuum chamber with a vacuum system or injecting helium directly into the equipment. A helium mass spectrometer is then used to detect any helium that may be present inside or outside the system being analyzed. If helium is detected where it should not be, it indicates a leak point. This method can quickly and accurately detect product leaks, thereby ensuring product quality and reliability. The working principle of sector field mass spectrometer is shown in Fig. 1.16.

Fig. 1.16 Working principle of sector field mass spectrometer

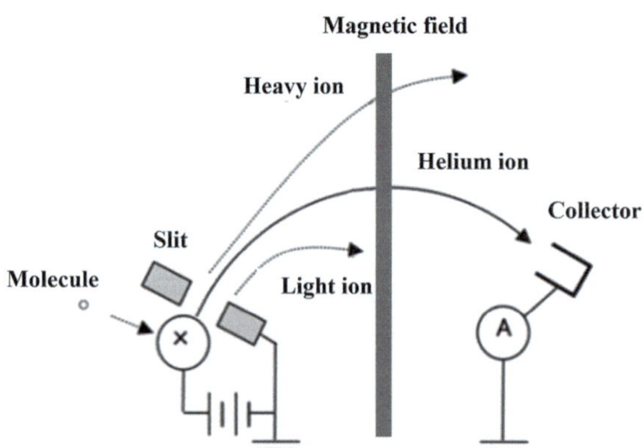

1.2.7 Breathing Mixture

Helium is increasingly valued in deep-sea diving applications. In the past, divers used compressed air to provide oxygen when diving to the seabed. However, due to the high pressure in deep-sea environments, the solubility of nitrogen in the blood increases with pressure, leading to nitrogen bubbles accumulating in the blood. When divers ascend too quickly, the pressure drops suddenly, causing dissolved nitrogen to run out, similar to opening a carbonated drink bottle and producing foam. This rapid ascent can trigger off "decompression sickness", where bubbles block blood vessels. Furthermore, the concentration of nitrogen increases with depth in deep water, and high nitrogen concentrations can affect the brain, causing a condition similar to alcohol intoxication, known as "nitrogen narcosis". To address these issues, helium–oxygen mixtures have gradually become preferred breathing gases by replacing nitrogen with helium. Later research and testing have found that adding a reasonable proportion of nitrogen to helium–oxygen mixtures can mitigate the High-Pressure Neurological Syndrome (HPNS) caused by helium. As a result, helium-nitrogen–oxygen mixtures have gained widespread acceptance.

In addition to deep-sea diving, helium-containing breathing mixtures are used in submarines. The use of helium in escape pods and life jackets provides better buoyancy, making escape processes safer and more efficient. The application of helium breathing mixtures in deep-sea diving is illustrated in Fig. 1.17.

Fig. 1.17 Helium breathing mixtures in deep-sea diving [23]

Column 1–5: The History of Helium–Oxygen/Helium-Nitrogen–Oxygen Mixtures in Diving

In 1919, the electrical engineer and inventor Elihu Thompson proposed that helium might be the best substitute for nitrogen in the breathing mixture used by divers. He suggested that the United States Bureau of Mines (USBM) conduct research to investigate the feasibility of using helium–oxygen mixtures in place of nitrogen. Starting in 1924, the United States Navy Department (USN) and the USBM jointly sponsored a series of research experiments on helium–oxygen mixtures. The results of these experiments demonstrated that helium–oxygen mixtures offered significant advantages over air for deep-sea diving.

Early helium–oxygen diving equipment. *Source* U.S. Navy Diving Manual 2016

In 1937, the United States Navy Research Laboratory (NRL) made its first attempt to switch the breathing gas for divers from air to a helium–oxygen mixture, in a bid to boost efficiency in deep-sea diving. However, when the divers ascended to a depth of 100 m and the breathing gas was switched back to air, they immediately experienced severe symptoms of "nitrogen narcosis". This marked the first time that the narcotic effect of nitrogen was observed.

By 1939, helium–oxygen mixtures were put into practical use. The U.S. Navy employed helium–oxygen mixtures as an air substitute during the salvage operation of the USS Squalus submarine, which was submerged at a depth of 73 m.

In 1957, George Foote Bond, hailed as the "Father of Saturation Diving", initiated the Genesis project, which proved that humans could endure prolonged exposure to different breathing mixtures in high-pressure environments.

In 1963, for the first time, a helium-nitrogen–oxygen mixture was used in a 6-day saturation dive at a depth of 30 m.

In 1965, diver Nic Flemming conducted a comparative test between air diving and nitrogen–oxygen mixture diving, discovering that nitrogen narcosis was only one of the many factors affecting divers' performance underwater.

In 1979, a research team led by Peter B. Bennet, the former president of Divers Alert Network (DAN), conducted the "Atlantis Series Dives", which analyzed the mechanisms behind using helium-nitrogen–oxygen mixtures to prevent High Pressure Nervous Syndrome (HPNS). The team concluded that adding an appropriate amount of nitrogen to the helium–oxygen mixture could mitigate the symptoms of HPNS caused by helium.

In 1983, cave diver Jochen Hasenmayer set a depth record of 212 m in Vaucluse, France, using a helium–oxygen mixture. In 1987, William C. Bill Stone led a team in exploring and mapping Wakulla Springs in Florida. This marked the first large-scale application of helium–oxygen and helium-nitrogen–oxygen mixtures in cave diving. The team successfully reached a depth of 1270 m using helium-nitrogen–oxygen mixtures.

Subsequently, Billy Deans, a pioneer in technical and wreck diving, and a former board member of the International Association of Nitrox and Technical Divers (IANTD), began teaching the use of helium-nitrogen–oxygen mixtures in recreational diving. Tom Mount, the former president of IANTD, developed the first training standards for helium-nitrogen–oxygen mixture courses, contributing to the widespread adoption of these gases.

In 2001, John Bennet set a Guinness World Record by reaching a depth of 305 m using a helium-nitrogen–oxygen mixture.

In 2014, Ahmed Gabr set the latest scuba diving depth record of 332.35 m in the Red Sea, Egypt, also using a helium-nitrogen-oxygen mixture.

1.3 Quality Standards for Helium Products

Different purity levels of helium are suited for various applications. Helium with 99% purity is primarily used to provide buoyancy in balloons and airships. Helium with 99.999% purity is commonly used in weather balloons, leak detection, welding protection, breathing mixtures, rockets, and medical applications. Helium with 99.9995% purity is often used in chromatographic analysis, optical fiber and semiconductor processing, laboratory research, and so forth. Helium with a purity of 99.9999% and above is primarily used in the manufacturing of semiconductor chips.

Helium gas contains a variety of impurities, and the quality standards for helium products specify the technical requirements for different specifications and impurity levels of helium products. Among various impurities, neon shares very similar physical and chemical properties with helium, such as low boiling point, small molecular size, and resistance to adsorption. Consequently, conventional methods tend to further

Table 1.4 Pure helium, high pure helium and ultra pure helium (**GB/T 4844-2011**)

Item		Indicator			
		Pure helium		High-pure helium	Ultra-pure helium
Helium (He) purity/% (by volume)	≥	99.99	99.995	99.999	99.9999
Neon (Ne) content	<	40	15	4	1
Hydrogen (H_2) content	<	7	3	1	0.1
Oxygen (O_2) + Argon (Ar) content	<	5	3	1	0.1
Nitrogen (N_2) content	<	25	10	2	0.1
Carbon monoxide (CO) content	<	1	1	0.5	0.1
Carbon dioxide (CO_2) content	<	1	1	0.5	0.1
Methane (CH_4) content	<	1	1	0.5	0.1
Water (H_2O) content	<	20	10	3	0.2
Total impurities content	≤	100	50	10	1

Note Unit is mL/m^3 (ppmV) unless otherwise specified

enrich neon in helium, leading to poor removal of neon from helium. Neon is thus the most challenging impurity to eliminate from helium. Therefore, in relevant quality standards, the upper limit for neon is allowed to be the highest, with its permissible impurity levels being more lenient compared to other elements.

1.3.1 Chinese Helium Product Standards

The Chinese national standard *Pure Helium, High Pure Helium and Ultra Pure Helium* (GB/T 4844-2011) specifies the technical requirements, inspection rules and methods, product packaging, labeling, storage, transportation, and safety warnings for helium. The specific technical requirements are detailed in Table 1.4.

1.3.2 U.S. Helium Product Standards

On August 1, 2014, the U.S. Department of Defense implemented the *Performance Specification-Propellant Pressurizing Agent, Helium* (MIL-PRF-27407D-2014), commonly referred to as the Military Standard. This specification defines the standards for helium products. The specific quality indicators are shown in Table 1.5.

The U.S. Compressed Gas Association (CGA) revised the *Commodity Specification for Helium* (CGA G-9.1-2018) standard in its 9th edition, which came into

Table 1.5 Grade limits for helium (gaseous only) **(MIL-PRF-27407D-2014)**

Item		Helium limits	
		Grade A	Grade B
Purity, % (by volume)	≥	99.995	99.997
Impurity	≤	50	–
Water		9	9
Hydrocarbons (as methane)		5	1
Oxygen		3	3
Nitrogen + Argon		14	5
Neon		23	23
Hydrogen		1	1
Carbon dioxide		1	–
Carbon monoxide		1	–

Note Unit is mL/m^3 (ppmV) unless otherwise specified

effect in 2018. This standard also specifies the quality indicators for helium, which are detailed in Table 1.6. CGA is a trade association for the U.S. industrial and medical gas supply industry, established in 1913 as a non-profit trade association and standards development organization dedicated to promoting safety standards and practices in the industrial, medical, and specialty gas industries.

1.4 Isotope Helium-3 (^{3}He)

^{3}He is a rare, non-radioactive isotope of helium. It is the only stable isotope with more protons than neutrons, apart from protium(^{1}H). ^{3}He has significant applications across various fields, including national security, medicine, industry, and scientific research. Following the "9/11" terrorist attacks in 2001, the U.S. federal government began deploying neutron detectors at borders to prevent the smuggling of nuclear and radioactive materials. This deployment led to a sharp increase in the demand for ^{3}He, resulting in a supply shortage. Moreover, Ouyang Ziyuan, the head of China's lunar exploration program and a renowned cosmochemist and geochemist from the CAS, has repeatedly stated that ^{3}He is an ideal material for the next generation of nuclear fusion power generation.

On Earth, ^{3}He is primarily primordial helium captured from the early solar system during the planet's formation, constituting only about 0.1% of the Earth's helium. Despite its scarcity, ^{3}He has become a focial point of research due to its unique properties, attracting the interest of experimental and theoretical physicists, physical chemists, and cryogenic engineers. However, because of its rarity and high cost, studying the properties of ^{3}He at extremely low temperatures is challenging but highly valuable. Since ^{3}He's properties are closely associated with the evolving quantum

Table 1.6 Directory of limiting characteristics of helium (**CGA G-9.1-2018**)

Quality verification level (grades)						
Limiting characteristics	Maxima for gaseous helium[a,b]					
	H	J	L	N	P	G
Helium/ % ≥	97.5	99.0	99.995	99.997	99.999	99.9999
Air/ %	–	1	–	–	–	–
Carbon dioxide	–	–	–	–	Sum = 0.5	Sum of all these impurities is less than 1 ppm
Carbon monoxide	–	10	–	–		
Hydrogen	–	–	–	1	1	
Neon	–	–	–	23	2	
Nitrogen + Argon	–	–	–	5[c]	5	
Odor	–	None	–	–	–	
Oxygen	–	–	5	3	1	
Total Hydrocarbon (as methane)	–	–	–	1	0.5	
Water Dew point °F °C	–	–	15 – 72 – 57.8	3 – 92 – 68.9	1.5 – 101 – 73.9	
Buoyancy	Yes	–	–	–	–	–
Identity	–	Yes	–	–	–	–

Note Unit is mL/m^3 (ppmV) unless otherwise specified
[a]These quality verification levels apply only to gaseous helium. Impurity limits for liquid helium are not specified since sufficient technical data and analytical procedures are not available to warrant a definitive quantitative specification. The requirement for ensuring that the loaded fluid in a container is liquid helium can be satisfied by analyzing the shipping container's vent gas or by demonstrating that the temperature of the loaded fluid is below the hydrogen triple point (13.8 K)
[b]Permanent particulates in the liquid can be reduced by using a filter assembly of 10 μm or less installed in the transfer line as the liquid enters the transport container
[c]Maximum 5 ppm nitrogen only

theory, research on ^{3}He has been at the forefront of global physics since the mid-twentieth century. One of the most remarkable discoveries was made by American scientist David Morris Lee and others in 1972, when they observed superfluidity in ^{3}He. This research was awarded the Nobel Prize in Physics in 1996.

The existence of ^{3}He was first proposed by Marcus Lawrence Elwin Oliphant and others in 1934. However, it was not until 1948 that large quantities of ^{3}He became available and the groundwork for large-scale experiments was laid, opening the way to related research. At the early stage, gaseous ^{3}He was considered a noble gas similar to ^{4}He, receiving little attention. As research into the liquefaction of ^{3}He progressed, perceptions began to change, leading to in-depth studies of liquid and even solid ^{3}He. The unique properties of liquid ^{3}He garnered widespread interest among scientists, and it came to be regarded as an ideal specimen of a Fermi quantum fluid, closely related to quantum mechanics. The study of ^{3}He has had a profound impact not only on experimental physics and cryogenic engineering but also on the development of theoretical physics and quantum mechanics.

In 1949, Darrell W. Osborne, Bernard Weinstock, and Bernard M. Abraham, among others, began pioneering research on liquid ^{3}He [5]. By 1951, Darrell W. Osborne and others solidified ^{3}He [18].

Chinese researchers have also conducted extensive studies on ^{3}He. For a long time, there was a lack of an equation of state for ^{3}He that covered the transition from liquid phase to gas phase at room temperature, across a wide range of pressures from low to tens of megapascals. In light of this, Huang Yonghua and others at Zhejiang University proposed the world's first wide-temperature-range, full-fluid, high-precision equation of state for ^{3}He, along with phase equilibrium equations for gas–liquid and solid–liquid states, as well as equations for transport properties such as thermal conductivity and viscosity. They also developed p–h (Pressure-enthalpy) and T-s (Temperature-entropy) diagrams required for refrigeration engineering, covering temperatures from 3 mK to 1500 K and pressures up to 20 MPa.

1.4.1 Physical Properties of ^{3}He

^{3}He and ^{4}He are both monoatomic gases with atomic nuclei containing two protons and two electrons orbiting outside the nucleus. The difference between them lies in the number of neutrons: ^{3}He has one neutron, giving it an atomic mass of 3.0162, while ^{4}He has two neutrons, with an atomic mass of 4.0026 [24].

^{3}He and ^{4}He are isotopes with similar physical and chemical properties, including being inert gases with extremely low critical temperatures, lacking a triple point, and being colorless, odorless, non-toxic, and non-flammable at room temperature and atmospheric pressure, and they also exhibit quantum fluid characteristics, such as superfluidity. However, due to slight differences in their atomic structures, their physical properties differ, especially at low temperatures.

From a quantum mechanics perspective, ^{3}He has an odd nuclear spin, making it a fermion, while ^{4}He has an even nuclear spin, making it a boson. Under low-temperature conditions approaching "absolute zero", both isotopes adhere to quantum mechanical principles but follow different statistical laws: ^{4}He obeys Bose–Einstein (BE) statistics and undergoes condensation at 2.178 K, transitioning into a superfluid state; ^{3}He follows Fermi–Dirac (FD) statistics and only undergoes a similar BCS superconducting condensation at 2.6 mK, transitioning into a superfluid state at a temperature three orders of magnitude lower than the λ transition temperature of ^{4}He. Furthermore, at the same temperature, ^{3}He has a much higher vapor pressure than ^{4}He—for example, at 1 K, the saturated vapor pressure of ^{3}He is 80 times greater than that of ^{4}He, and at 0.5 K, the difference is nearly 10,000 times. Due to its non-zero nuclear magnetic moment, ^{3}He has a greater spin entropy than ^{4}He, resulting in a higher entropy value at low temperatures. For instance, at 1 K, the specific heat capacity of ^{3}He is 40 times greater than that of ^{4}He, with this difference increasing as the temperature decreases.

Table 1.7 Main thermodynamic properties of ^{3}**He**

Parameter	Value	Parameter	Value
Atomic mass	3.0162	Magnetic moment (nuclear magnetons)	– 2.127
Normal boiling point/K	3.19	Density at 0 K/(kg/m^3)	82.3
Density at room temperature (21.1 °C)/(kg/m^3)	0.1650	Compressibility at 0 K/(mm^3/J)	361
Critical temperature Tc/K	3.3157	Latent heat of vaporization at 0 K/(J/mol)	20.56
Critical pressure p_c/MPa	0.1146	Surface tension at 0 K/(N/m)	0.16×10^{-4}
Critical density ρ_c/(kg/m^3)	41.19	Speed of sound at 0 K/(m/s)	183
Lowest melting pressure/MPa	2.9	Specific heat capacity at 1.0 K/[J/(mol·K)]	4.2
Temperature at lowest melting pressure/K	0.3152	Vapor pressure at 1.7 K/Pa	11
Superfluid phase transition temperature/mK	2.6	Thermal conductivity at 3.2 K/[mW/(m·K)]	20.000
Triple point	–	Viscosity at 3.2 K/Pa.s	1.9×10^{-6}

The main thermodynamic properties of ^{3}He are summarized in Table 1.7.

1.4.2 *Extraction of* ^{3}*He*

In the 1950s, the production of ^{3}He was extremely limited. Nevertheless, the situation has since improved significantly, and ^{3}He is now available through commercial channels. The book *Helium-4 and Helium-3 and Their Applications*, authored by Zhang Peng, Huang Yonghua, and Chen Guobang, provides a detailed explanation of the extraction methods and applications of ^{3}He [24, 25]. There are three main methods to obtain ^{3}He.

1. Neutron Capture Reaction of ^{6}Li

Since the mid-1940s, the primary method for obtaining ^{3}He has been through the neutron capture reaction of lithium-6 (^{6}Li). This method is relatively simple and reliable, and experimental research on ^{3}He began after it was discovered that ^{3}He was a byproduct of the ^{6}Li neutron capture reaction.

$$ {}_{3}^{6}\mathrm{Li} + {}_{0}^{1}\mathrm{n} \rightarrow {}_{1}^{3}\mathrm{H} + {}_{2}^{4}\mathrm{He} \quad (1.4) $$

$$ {}_{1}^{3}\mathrm{H} \rightarrow {}_{2}^{3}\mathrm{He} + \beta \quad (1.5) $$

According to Eqs. (1.4) and (1.5), it can be understood that when ^6Li captures a neutron, it produces tritium (^3H), a radioactive isotope of hydrogen, and ^4He. Subsequently, the mixture of tritium and ^4He generated by heating inside a palladium tube undergoes a separation process where tritium can pass through the palladium tube while ^4He cannot. This property allows for the isolation of tritium from the mixture. However, the separated tritium is not stable, which undergoes β decay and transforms into ^3He.

Given the relatively abundant quantity of ^6Li in nature, using large quantities of neutrons provided by experimental nuclear reactors to act on ^6Li to produce ^3He is highly feasible. This method yields more ^3He and at a lower cost than separating it from natural helium. Moreover, since this method essentially relies on the β decay of tritium to produce ^3He, seawater, which contains a significant amount of tritium, can be considered an important source. It is imperative to make a breakthrough in technology for extracting tritium from seawater as soon as possible, which could further reduce the difficulty and cost of acquiring ^3He.

2. Low-Temperature Separation of $^3He/^4He$

The most direct way to obtain ^3He is by separating it from natural helium (a mixture of ^4He and ^3He). Air, particularly natural gas, contains a certain amount of helium, which can be obtained through cryogenic gas separation technologies. Nevertheless, the concentration of ^3He in the resulting helium is extremely low. For instance, in helium obtained from air, the ratio of ^3He to ^4He is approximately 1.4×10^{-6}:1. The concentration difference is even greater in helium obtained from natural gas. Therefore, under such low concentrations, obtaining large amounts of ^3He through cryogenic gas separation technologies still requires further research.

3. The Moon—A Major Source of ^3He in the Future

In September 1986, researchers from the University of Wisconsin-Madison, including Layton J. Wittenberg, confirmed the presence of large amounts of ^3He in the lunar surface soil after analyzing lunar soil sample data [26]. The solar wind, a stream of high-energy particles emitted by the Sun, contains a significant amount of ^3He. Since the Moon lacks an atmosphere, ^3He from the solar wind directly accumulates on the lunar surface over time, becoming an important component of lunar soil. According to related data, the total amount of ^3He on the Moon could reach 1×10^6 to 5×10^6 tons.

The technological route in Fig. 1.18 depicts how to extract ^3He from the Moon. First of all, solar energy is used to heat lunar soil to 300 K to separate out fuel hydrogen. Then, by using a radiator and condenser, the temperature is lowered to 50 K to obtain essential life-supporting substances like water, carbon dioxide, and nitrogen. Finally, the temperature is further reduced to 1.5 K for isotopic separation to obtain ^3He. Notably, for every ton of ^3He obtained from the Moon, 6100 tons of H_2, 3100 tons of ^4He, and 3300 tons of H_2O can be simultaneously produced as valuable byproducts. While the current challenges in traveling to and from the Moon, such as high costs and reliability issues, are significant, solving these problems and establishing a ^3He

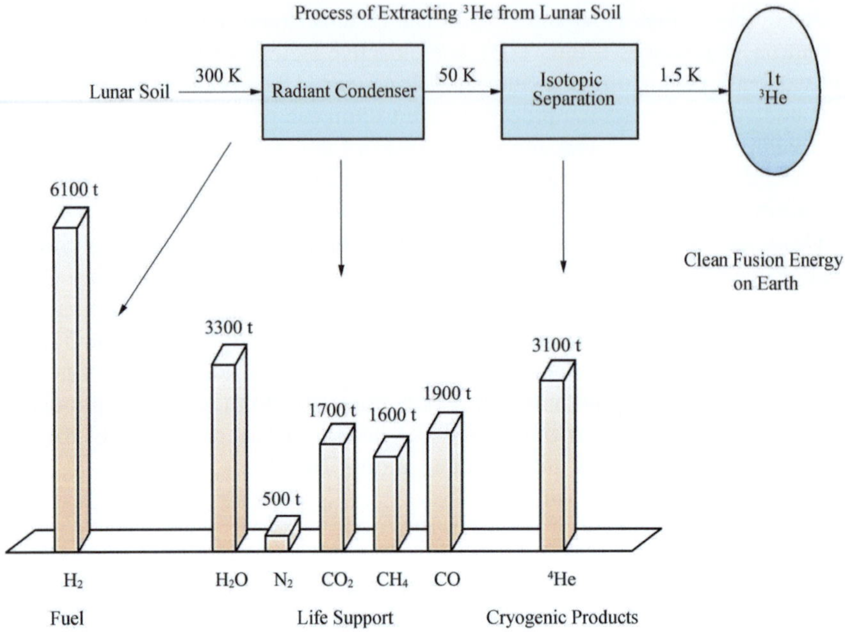

Fig. 1.18 Technological route for extracting ^{3}He from the moon

processing base on the Moon could yield significant benefits. This potential might partly explain why countries like the United States, Russia, China, Japan, Israel, and India have been competing to initiate a new wave of lunar exploration since the 1990s. In 2007, the launch of China's Chang'e-1 satellite reignited the search for ^{3}He. After more than a year of analyzing microwave data from the Chang'e-1 lunar satellite, it was estimated that the lunar soil contains about 1×10^6 tons of ^{3}He, which, when converted into nuclear energy, could power the Earth for 10,000 years.

1.4.3 Typical Applications of ^{3}He

1. Dilution Refrigeration

^{3}He possesses properties such as low boiling point, low density, high specific heat capacity, and high thermal conductivity, which make it highly valuable in achieving ultra-low temperatures below 1 K. This temperature range is of great importance to fundamental physics research and other modern high-tech scientific endeavors, thus drawing significant interest from experts in low-temperature physics and cryogenic engineering.

In 1956, Walters and Fairbanks discovered that when the temperature drops below 0.87 K, a mixture of ^{3}He and ^{4}He separates into two immiscible phases: a lighter

^{3}He-rich phase floating on top and a heavier ^{4}He-rich phase at the bottom. Below 0.3 K, the ^{3}He-rich phase is almost pure ^{3}He, while the ^{4}He-rich phase contains about 6.4% ^{3}He—a proportion that remains constant even near "absolute zero". This unique property of ^{3}He forms the basis for dilution refrigerators, which can continuously achieve millikelvin temperatures. In 1962, Heinz London and Kurt Mendelssohn, among others, proposed a practical scheme for dilution refrigeration.

At low temperatures, ^{4}He exhibits superfluidity, is an inert liquid, while ^{3}He remains a normal fluid, acting as the active component. When a container holds a mixture of ^{3}He and ^{4}He, the ^{4}He-rich phase at the bottom supports the ^{3}He-rich phase above it. If the dissolved ^{3}He in the ^{4}He-rich phase is gradually removed to lower its concentration, the equilibrium between the two phases will be disrupted, leading to the diffusion of ^{3}He atoms from the ^{3}He-rich phase across the boundary layer into the ^{4}He-rich phase. From an interface perspective, this process resembles the "evaporation" of ^{3}He into the superfluid ^{4}He phase, akin to evaporative refrigeration. As a matter of fact, this process represents a continuous dilution of ^{3}He. If this dilution persists, the liquid is continuously cooled, hence the term "dilution refrigeration" [24].

In ^{3}He evaporative refrigeration, as the temperature decreases, the vapor pressure of ^{3}He drops sharply, resulting in the inability to sustain the refrigeration effect. Therefore, the minimum temperature achievable by this approach is 0.25 K. In contrast, ^{3}He/^{4}He dilution refrigeration stands unaffected by the vapor pressure of ^{3}He. The ^{3}He content in the ^{4}He-rich phase remains constant, and a vacuum pump can maintain a stable ^{3}He circulation, allowing for lower temperatures than those achievable by ^{3}He evaporative refrigeration. This is also the most significant difference between the two cooling methods.

2. Fusion Energy Resource

^{3}He is an ideal material for fusion reactions. In fusion reactors using deuterium (^{2}H) and ^{3}He as fuel, clean and safe fusion energy can be produced. It is estimated that only 100 tons of ^{3}He would be sufficient to meet the global electricity demand for one year. The fusion products of this reaction are non-radioactive, with minimal neutron power, decay heat, activation radioactivity, and material radiation damage. The neutron power constitutes approximately 5% of the fusion power, which can reduce the radiation shielding requirements for the reactor and increase component life and magnetic field utilization. Moreover, the entire reaction process releases a tremendous amount of energy.

3. Production of Tritium (^{3}H)

^{3}H is of irreplaceable importance in the nuclear field, primarily used to manufacture hydrogen bombs and boost the utilization of fission materials in atomic bombs. This helps increase the explosive yield of fission and thermonuclear weapons, making them more compact.

The primary method of producing tritium (^{3}H) is through the neutron capture reaction of lithium-6 (^{6}Li). ^{3}H undergoes β decay to generate ^{3}He, and this is currently the main source of ^{3}He. Due to its radioactivity and instability, the ^{3}H currently used in nuclear weapons is mainly produced artificially through the reaction between neutrons and ^{6}Li; it can also be generated via the reaction between neutrons and ^{3}He.

4. Nuclear Magnetic Resonance

The nuclear spin of ^{3}He is odd, giving it a nuclear magnetic moment, which enables it to exhibit NMR in a magnetic field. Laser-enhanced polarized ^{3}He is an ideal sample for magnetic resonance imaging (MRI), offering greater advantages over conventional xenon-129 (^{129}Xe) MRI. ^{3}He has a magnetic moment 2.7 times greater than that of ^{129}Xe, resulting in a stronger magnetic resonance signal at a given polarization density. Additionally, ^{3}He provides higher relative sensitivity, signal stability, and better imaging quality.

5. Research on Solar Characteristics

Thermonuclear reactions in the Sun produce energy through nuclear fusion, involving the generation of large quantities of ^{3}He particles. By analyzing the characteristics of ^{3}He in the solar spectrum, we can indirectly study the Sun's internal behavior. Since the first discovery in 1962 of significant ^{3}He enrichment in solar high-energy particles, scientists have extensively studied solar ^{3}He enrichment from experimental observations and theoretical models. These studies have provided valuable insights into the Sun's operation and development.

6. Neutron Detector

Over the past 20 years, neutron detection technology has seen continuous development and refinement, particularly in studying neutron structure, dynamics, composites, the magnetization of condensed matter, and the development of high-intensity neutron sources. Neutron detection technology plays a critical role in nuclear medicine and clinical diagnostics, nuclear power plant safety inspections, space physics, aerospace, and industrial applications. Due to its high sensitivity, ^{3}He is often used as the detection medium in neutron detectors.

Since the 9/11 terrorist attacks in 2001, the U.S. Department of Homeland Security has widely deployed neutron proportional counters at borders and throughout the country to prevent the smuggling of plutonium or other weapon materials. Meeting the demand for these counters requires about 2×10^{4} L of ^{3}He annually.

Column 1.6: Planetary Engine

^{3}He is an ideal fusion reaction material with extremely high energy density. The energy released per unit mass of ^{3}He is millions of times that of oil and more than five times that of the hydrogen isotopes deuterium-tritium (D-T) fusion reaction. Additionally, the thermal nuclear reactions of ^{3}He do not produce neutrons during operation (^{3}He undergoes thermal nuclear reactions with deuterium, producing only non-radioactive protons as products). In the widely acclaimed science fiction film The Wandering Earth, the planetary engine utilizes vacuum rock nuclear fusion technology as its power source, employing ^{3}He to initiate a silicon atomic nucleus fusion chain reaction. This process generates high-energy, high-pressure plasma streams, which propel the Earth out of the solar system.

References

1. Wheeler M. Helium: the disappearing element. New York: Springer, New York; 2015.
2. Vollhardt D, Wolfle P. The superfluid phases of helium 3. New York: Courier Corporation; 2013.
3. Oliphant MLE, Harteck P, Rutherford E. Transmutation effects observed with heavy hydrogen. Proc R Soc Lond. 1934;144(853):692–703.
4. Van Sciver SW, Timmerhaus KD, Clark AF. Helium cryogenics. New York: Springer, New York; 2012.
5. Balibar S. Laszlo tisza and the two-fluid model of superfluidity. C R Phys. 2017;18(9–10):586–591.
6. Dong X, Oganov AR, Goncharov AF, et al. A stable compound of helium and sodium at high pressure. Nat Chem. 2017;9:440–445.
7. Weisend II JG. Superfluid: how a quantum fluid revolutionized modern science. Cham: Springer International Publishing; 2023. pp. 35–42.
8. Zhang P, Wang RZ. Heat transfer of superfluid helium. Beijing: Science Press; 2009 (in Chinese).
9. Chinese Association of Refrigeration. Research report on strategic development of China's refrigeration industry. Beijing: China Architecture & Building Press; 2016 (in Chinese).
10. Ballentine CJ, Burnard PG. Production, release and transport of noble gases in the continental crust. Rev Mineral Geochem. 2002;47(1):481–538.
11. Wang XB. Geochemistry and cosmochemistry of noble gas isotopes. Beijing: Science Press; 1989 (in Chinese).
12. Kaneoka I, Takaoka N. Noble-gas state in the earth's interior—some constraints on the present state. Chem Geol Isot Geosci Sect. 1985;52(1):75–95.
13. Mamyrin BA, Tolstikhin IN. Helium isotopes in nature. Amsterdam: Elsevier; 2013.
14. Xu G, Li YH, Wang ZQ, et al. Progress of investigation and evaluation of helium resources in China. Acta Geol Sin. 2023;97(5):1711–1716 (in Chinese).
15. Arkharov AM. Helium: history of its discovery, technology of its liquefaction, areas of its application. Chem Pet Eng. 1995;31:50–60.
16. Arkharov AM. Achievements and problems of engineering cryology. Chem Pet Eng. 2007;43(7):389–399.

17. Alvarez LW. Discovering alvarez: selected works of Luis W. Alvarez with commentary by his students and colleagues. Chicago: University of Chicago Press; 1987.
18. Osborne DW, Abraham BM, Weinstock B. Solidification of He^3. Phys Rev. 1951;82(2):263–264.
19. Sasaki K, Li HW, Hayashi A, et al. Hydrogen energy engineering: a Japanese perspective. Stuttgart, Germany: Springer Nature; 2016. pp. 53–92.
20. Liu L, Tian S, Xue D, et al. Industrial feedforward control technology: a review. J Intell Manuf. 2019;30:2819–2833.
21. Vaamonde Couso E, Vázquez Gómez J. Laser beam welding and automotive engineering. In: Structural connections for lightweight metallic structures. Berlin, Heidelberg: Springer Berlin Heidelberg; 2011. pp. 59–84.
22. Tian ZY, Mbayachi VB, Dai WK, et al. Gas chromatography/mass spectrometry. advanced diagnostics in combustion science. Singapore: Springer Nature Singapore; 2023. pp. 33–69.
23. Cohen D. Transformative potential of sustainable Scuba Dive tourism. In: Sustainable development and resilience of tourism: wellbeing and quality of life perspectives. Cham: Springer International Publishing; 2024. pp. 121–150.
24. Huang YH. Research on the equation of state and thermophysical properties of Helium-3. Hangzhou: Zhejiang University; 2006 (in Chinese).
25. Zhang P, Huang YH, Chen GB. Helium-4 and Helium-3 and their applications. Beijing: National Defense Industry Press; 2006 (in Chinese).
26. Wittenberg LJ, Santarius JF, Kulcinski GL. Lunar source of ^3He for commercial fusion power. Fusion Technol. 1986;10(2):167–178.

Chapter 2
Helium Resources and Market

Yingke Li and Yafei Zhang

The global distribution of helium resources is uneven, with significant differences in resource quality across regions. Currently, the majority of the world's helium production capacity is concentrated in a few countries. As helium finds increasing applications across various fields, global demand for helium is expected to continue growing. The helium market is generally characterized by a supply–demand imbalance, with supply often unable to keep pace with demand. Large helium extraction projects, typically impact the market for about 3–5 years, creating strong market fluctuations. China has relatively scarce helium resources with lower resource quality. As a result, the country has yet to establish a robust industrial chain of helium, whose consumption largely depends on imports. In recent years, however, as domestic helium extraction projects entered the production phase, the production capacity has witnessed a rapid growth, providing strong support to ease the supply constraints. Looking forward, it is imperative to build a stable and resilient industrial chain, reduce reliance on imports, and ensure the security of the national helium supply.

This chapter provides an overview of the distribution of helium resources in China and other countries, the global helium market's supply and demand dynamics, and the supply–demand situation in the Chinese helium market. It also presents our vision and directions from the perspective of strategically ensuring a stable supply.

Y. Li (✉)
China Petroleum Engineering & Construction Corporation Southwest Company, Chengdu, China
e-mail: liyingke_sw@cnpc.com.cn

Y. Zhang
China Petroleum Engineering & Construction Corporation Southwest Company, Chengdu, China
e-mail: zhangyafei_sw@cnpc.com.cn

L. Pu (ed.), *Helium Engineering Technologies*,
https://doi.org/10.1007/978-981-96-4086-7_2

2.1 Distribution of Helium Resources

Globally, helium resources are distributed unevenly, primarily concentrated in the United States, Qatar, Algeria, Russia, and Iran, which together account for over 80% of the world's helium resources. Moreover, the quality of helium resources varies significantly across different regions, with varying average helium content in natural gas fields. Recent exploration studies have shown that China has some potential helium resources, but compared to the global level, China's helium resources are relatively scarce and of lower quality.

2.1.1 Global Distribution of Helium Resources

The world's identified and developed helium resources are mainly found as free, trace components in oil and gas reservoirs or non-hydrocarbon gas fields, with an uneven distribution on the whole. According to the 2023 statistics from the United States Geological Survey (USGS), the global helium resources are estimated at approximately 5.19×10^{10} m^3, with the majority of resources concentrated in the United States (2.06×10^{10} m^3), Qatar (1.01×10^{10} m^3), Algeria (8.2×10^9 m^3), Russia (6.8×10^9 m^3), Canada (2.0×10^9 m^3), and China (1.1×10^9 m^3), accounting for about 94.03% of the global total [1], with absolute resource advantages, as shown in Fig. 2.1.

Helium resources are unevenly distributed not only across the globe but also within individual countries. For example, in the United States, as much as 97% of helium reserves are concentrated in six major gas fields: Cliffside, Hugoton, and Panoma, among others [2, 3]. In Russia, helium reserves are primarily located in the Far East gas fields, such as Kovykta, Chayanda, Sobin, and Srednebotuobin [4].

Additionally, the helium content in natural gas fields around the world varies significantly. While there are no universally accepted standards for helium content classification and grade evaluation, researchers have conducted extensive studies on the subject. Table 2.1 presents one classification and grade evaluation standard for

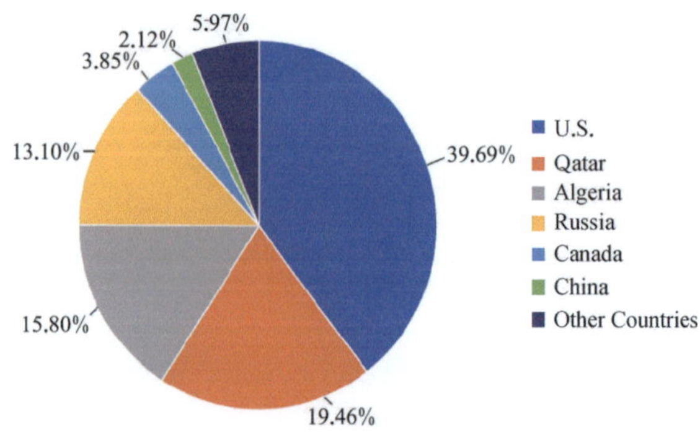

Fig. 2.1 Global helium resource distribution proportions [1]

Table 2.1 Helium content classification and grading evaluation standards

Classification		Helium content	
		% (by volume)	mL/m³ (ppmV)
Helium-rich reservoirs	Super-high helium content	> 10	≥ 100,000
	Ultra-high helium content	1.0–10	10,000–100,000
	High helium content	0.1–1.0	1000–10,000
Helium-poor gas reservoirs	Medium helium content	0.05–0.1	500–1000
	Low helium content	0.01–0.05	100–500
	Trace helium content	< 0.01	< 100

Table 2.2 Industrial classification standards for helium gas fields

Helium gas reserves/ 10^6 m³	Scale classification	Helium content/% (by volume)	Content classification
≥ 100	Ultra-large gas field	≥ 0.500	Ultra-helium-rich field
50–100	Large gas field	0.150–0.500	Helium-rich field
25–50	Medium gas field	0.050–0.150	Helium-containing gas field
5–25	Small gas field	0.005–0.050	Helium-poor gas field
< 5	Ultra-small gas field	< 0.005	Ultra-helium-poor gas field

helium content, and Table 2.2 provides the industrial classification of helium gas fields based on the 1984 VP Yakucheni (В. П. Якучени) standard [5, 6]. In the United States, which enjoys the richest helium resources in the world, the average helium content in natural gas fields in the eastern regions is about 0.8%, with some fields holding as much as 7.5% helium.

2.1.2 *Distribution of Helium Resources in China*

According to data released by the USGS, China accounts for only 2.12% of the global helium resources. Although this figure differs slightly from that provided by Chinese authoritative institutions, the discrepancy is minimal. Helium exploration in China began relatively late, with initial efforts focusing more on comprehensive surveys and less on helium-specific drilling. Recent research findings indicate that helium resources in China are scattered and exhibit characteristics of zonal and regional enrichment, suggesting a certain level of resource potential on the whole. Current exploration results show the presence of helium-containing natural gas in the Weihe Basin, Sichuan Basin, Tarim Basin, Qaidam Basin, Songliao Basin, Bohai Bay Basin, Subei Basin, Hailar Basin, and Ordos Basin, among others.

In large superimposed basins like the Tarim Basin, Sichuan Basin, Ordos Basin, and Qaidam Basin, multiple helium-rich natural gas reservoirs have been discovered along the uplifted edges. This aligns with international exploration experience, as globally, most large-scale helium-rich natural gas reservoirs with industrial exploitation value are located in intracontinental and marginal uplift zones. Intracontinental rift basins are also hotspots for global helium exploration. Moreover, significant helium-rich natural gas accumulations have been found around deep major faults, especially in non-hydrocarbon gas fields dominated by carbon dioxide and nitrogen. In China's eastern regions, several non-hydrocarbon helium-rich gas reservoirs have been identified around the Tan-Lu Fault Zone in the Songliao Basin, Bohai Bay Basin, and Subei Basin. Despite helium abundance, these reservoirs are generally small in scale, hence their limited resource potential [7].

Helium resources in China are primarily concentrated in large superimposed basins and intracontinental rift basins in the central and western regions. Preliminary forecasts suggest that the Sichuan Basin, Ordos Basin, Tarim Basin, Qaidam Basin, and Fenwei Basin have significant potential helium reserves, making them key exploration areas for rapidly boosting domestic production capacity [7].

The Weiyuan Gas Field in the Sichuan Basin is the first gas field in China to achieve commercial utilization of helium, with a helium content of approximately 0.2%, reaching a maximum of 0.36% [8].

The Ordos Basin is currently the largest natural gas production base in China, with helium content ranging from 0.03% to 0.05%. In the northern part of the basin, some gas fields have helium content exceeding 0.1%. The Dongsheng Gas Field, for instance, has a helium content ranging from 0.045% to 0.487%, with proven reserves of 2.444×10^8 m^3, controlled reserves of 4.270×10^8 m^3, and estimated reserves of 1.59×10^8 m^3, totaling 8.304×10^8 m^3 of geological reserves [9].

The southwestern, northern, and central regions of the Tarim Basin also show promising helium enrichment, with helium content ranging from 0.05% to 2.19% [10, 11]. The Hetianhe Gas Field on the southeastern edge of the Bachu Uplift has helium content ranging from 0.26% to 0.53%, attributed to crustal sources, with proven helium reserves of 1.9591×10^8 [12].

In the northern margin of the Qaidam Basin, high-grade helium-rich natural gas reservoirs have been successively discovered in the Dongping, Mabei, and Jianbei gas fields. The Dongping Gas Field has a helium content ranging from 0.075% to 1.069%, the Mabei Gas Field from 0.06% to 0.20%, and the Quanjishan area in the northern margin of the Qaidam Basin has helium content ranging from 0.21% to 0.72%, with standard helium resources estimated to reach 4×10^8 m^3 [13–16].

The Fenwei Basin has well-developed helium source rocks with high quality and substantial resource potential [17, 18]. In the mid-section of Xi'an Depression and Gushi Depression, many Cenozoic geothermal wells commonly exhibit associated crustal-derived helium, with 77 wells showing helium presence. Among these, 38 geothermal wells have gas with helium content greater than 1%, with the highest clocking up 9.226%. In the northern Jinzhong Depression, six new geothermal wells have been discovered with helium content exceeding 10%. The volume of water-soluble helium resource at a burial depth of less than 4000 m in the Weihe Basin is

2.13×10^9 m^3, with average water-soluble helium contents of 2.211%, 1.783%, and 1.077% in the Weinan, Xi'an, and Xianyang regions, respectively [19].

2.2 Global Helium Market

Helium possesses unique physical and chemical properties, making it an irreplaceable essential material in fields such as aerospace, healthcare, national defense, superconductivity, and high-tech manufacturing. It plays a critical role in advancing cutting-edge scientific research, the national defense industry, and high-tech industries. As a scarce strategic resource, the helium market has long been limited in scale. Currently, global annual helium demand is approximately 2×10^8 m^3, while annual production is only about 1.6×10^8 m^3, resulting in a significant supply gap. With the successive commissioning of large helium extraction plants, the market may experience substantial impacts, potentially leading to an oversupply of helium.

2.2.1 Global Helium Production

From the late twentieth century to the early twenty-first century, global annual helium production increased from 0.85×10^8 m^3 to 1.75×10^8 m^3, showing a general trend of steady growth. Subsequently, production levels stabilized, with annual output remaining around 1.6×10^8 m^3 from 2012 to 2022 [1], as shown in Fig. 2.2.

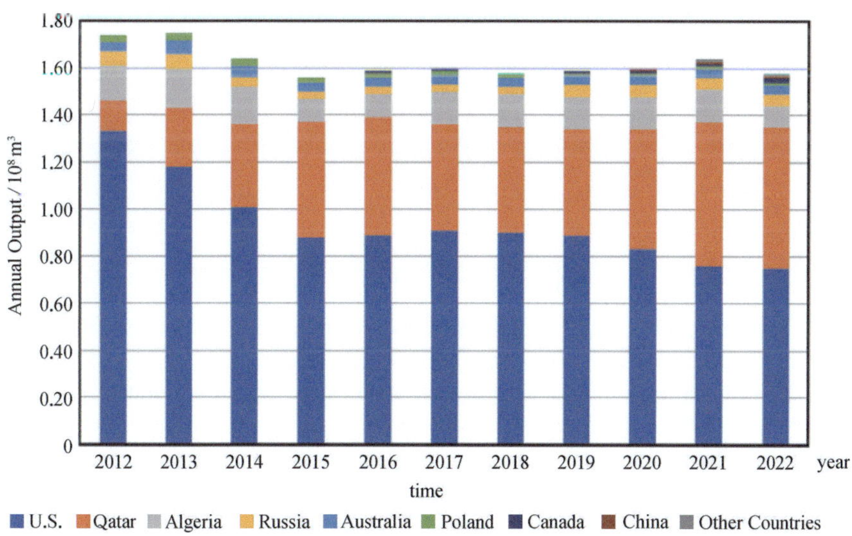

Fig. 2.2 Global helium production from 2012 to 2022

Global helium production is concentrated in a few countries or regions with abundant helium resources and high resource grades. The United States and Qatar hold the highest shares of global helium resources, dominating global production. From 2012 to 2022, the United States and Qatar together accounted for nearly 85% of global helium production, followed by Algeria and Russia, which contributed about 11% of global production.

2.2.2 Global Helium Demand

In recent years, with the widespread application of helium in numerous fields, global demand has been growing at an annual rate of approximately 5%. The market has generally been characterized by demand outstripping supply. Nevertheless, the commissioning of large helium extraction projects often causes market shocks lasting 3–5 years, leading to strong market volatility. Regionally, North America is the largest consumer of helium, accounting for about 42% of global consumption, making it the dominant consumption region. The Asia–Pacific region follows, with approximately 32% of global consumption, and Europe ranks third with about 20%. The Middle East and Africa account for only 5% of global helium consumption, indicating relatively low demand in these regions.

Currently, the use of helium in aerospace, healthcare, national defense, superconductivity, and high-tech manufacturing is becoming increasingly widespread, and demand continues to expand. With the continued development of these fields, helium demand on the whole is expected to keep growing. According to relevant institutional forecasts, global helium demand will reach 2.2×10^8 to 3.0×10^8 m^3 by 2030. Without new helium resource exploration and development to provide sustained support, global helium production could decline to 1.34×10^8 m^3, resulting in a supply–demand gap of up to 1.66×10^8 m^3 [7].

2.2.3 Global Helium Supply Chain

The United States, with its abundant and high-quality helium resources, was the first country to develop helium and has now become the world's leading helium producer. The U.S. began producing helium in the early twentieth century, and its production has consistently ranked at the top since then, which accounted for nearly 80% of the global helium supply in 2012. Qatar is the second-largest helium producer after the United States, with helium resources second only to the U.S. Although the helium content in Qatar's natural gas is low, the country capitalizes on the liquefied natural gas (LNG) production process to extract helium from boil-off gas (BOG), where helium concentrations can reach 1% to 2%, making it commercially viable. Qatar's helium extraction equipment and technology are mainly sourced from Europe and the U.S., and its sales are influenced by Western markets.

The top three helium suppliers—Linde PLC, Air Liquide S.A., and Air Products and Chemicals, Inc.—control 70% of the global market, giving them substantial influence in the industry. Linde PLC operates helium extraction plants in Kansas, Algeria, Qatar, and Australia, among others, and maintains 50 helium transshipment stations. In early 2020, Linde divested 10% of its market share to China's Guangzhou Guanggang Gases & Energy Co., Ltd., including helium contracts, sources, customers, and tank containers, reducing its market share from 40% to around 30%. Air Liquide is the largest buyer of Qatar's helium resources, with a helium storage facility and a well-established helium logistics system, holding approximately 20% of the market. Air Products and Chemicals, Inc. has significant advantages in the technology of cryogenic helium extraction from natural gas and the manufacturing of helium storage and transportation equipment, accounting for about 10% of the market.

Over the past decade, the global helium market has repeatedly faced supply shortages, drawing increased attention to helium production. Countries have been actively tapping into helium resource potential and deploying helium industries. Currently, there are 19 large-scale helium production plants worldwide, including 15 in the U.S., one in Qatar, one in Algeria, one in Russia, and one in Poland. Several new helium production plants are also being commissioned, such as the Amur Helium Extraction Plant and the Irkutsk Helium Extraction Plant, which were put into operation between 2021 and 2023. These developments could lead to significant shifts in the global helium supply landscape. As production increases, the helium market may experience oversupply, resulting in a rapid short-term price decline. According to relevant statistics, by the end of 2025, helium production capacity is expected to increase by 1.2×10^8 m^3, bringing the global helium supply to 2.6×10^8 m^3 (Table 2.3).

Table 2.3 Global new and planned helium extraction plants in **2021–2025** (partial) [7]

Country	Owner/operator	Plant/gas field	Helium capacity/(10^4 m^3/a)
Russia	Gazprom	Amur phase 1	2000
		Amur phase 2	2000
		Amur phase 3	2000
	Irkutsk Oil Company (INK)	Yarakta gas field	750
	Irkutsk Oil Company (INK)	Markovsk gas field	450
Qatar	Qatar energy, etc.	Ras Laffan-3	1200
Tanzania	Helium one	Helium One	2800
Algeria	Sonatrach, etc.	Algeria helium extraction plant	1000

2.3 China's Helium Market

China possesses a certain amount of helium resources, but these are generally of low quality, with most of the helium-containing natural gas fields containing poor helium concentrations. The high development costs have resulted in low domestic helium production. As a major consumer of helium, China primarily relies on imports to meet its demand.

2.3.1 *China's Helium Production Capacity*

In recent years, with progress in technologies, heightened awareness of resource reserves, and continued growth in downstream demand, China's helium extraction projects have gradually entered the production phase, resulting in a rapid increase in production capacity, particularly over the past two years. This growth has provided some relief to the supply constraints.

China's natural gas industry began experimenting with helium extraction in the 1960s, with the first helium extraction unit in Weiyuan handling 5×10^4 m^3/d of natural gas, yielding approximately 2×10^4 m^3/a of helium. In 2012, a new helium extraction facility was built in Dongxing Town, Rongxian County, Sichuan Province, with a production capacity of 2.1×10^5 m^3/a [20]. In recent years, driven by global resource protection measures and the growth of downstream industries such as semiconductors, China has accelerated the development of its helium industry. Multiple helium extraction projects have gradually entered production, contributing to a significant increase in helium production capacity by 2020. With the rapid development of China's BOG helium extraction technology and capacity expansion, helium output reached 1.30×10^6 m^3 in 2021 [7] and 3.258×10^6 m^3 in 2022.

China's helium production capacity is primarily concentrated in regions rich in natural gas resources, such as Inner Mongolia, Shaanxi, and Xinjiang. In 2022, Inner Mongolia accounted for 76.73% of the country's helium production capacity, making it the top producer, followed by Shaanxi at 10%. Ningxia and Chongqing each accounted for 3%, Gansu for 2%, and Shanghai, Hebei, and Hubei for 1% each (see Table 2.4). In 2023, Xinjiang's production capacity saw a rapid increase.

2.3.2 *Helium Imports*

China's helium consumption primarily relies on imports, with import volumes sustaining growth in recent years. In 2014, China imported 1.513×10^7 m^3 of helium. By 2021, helium imports had risen to 2.065×10^7 m^3, as shown in Fig. 2.3. China's helium imports are concentrated among a few sources, mainly Qatar, the United States, and Australia, as shown in Fig. 2.4. In 2022, 83.66% of China's

Table 2.4 Statistics of helium production enterprises in China (2022)

Province	Enterprise	Production capacity/(10^4 m^3/a)	Commissioning year
Sichuan	Chengdu Natural Gas Chemical Works	5	–
Inner Mongolia	Inner Mongolia Xingsheng Natural Gas Co., Ltd.	70	August 2020
Inner Mongolia	Inner Mongolia Sentai Natural Gas Co., Ltd.	7	July 2021
Inner Mongolia	Inner Mongolia Wanrui Natural Gas Co., Ltd.	75	December 2021
Inner Mongolia	Shuifa Energas Gas Co., Ltd.	23	Late 2021
Inner Mongolia	Inner Mongolia Zhengtai Yida New Energy Co., Ltd.	45	November 2022
Ningxia	China Energy North Natural Gas Co., Ltd.	11.2	August 2020
Hebei	Hebei Hansteel Gas	3	2015 and 2017
Hebei	Wuhan Iron and Steel Group Gas Co., Ltd.	2	2006 and 2018
Shanghai	Shanghai Baosteel Gases Co., Ltd.	1.5	Late 2019
Shanghai	Shanghai Qiyuan Air Separation Technology Development Co., Ltd.	3	Late 2015
Shaanxi	Yulin Yuanheng Energy Co., Ltd.	15	February 2022
Shaanxi	Shaanxi Yanchang Petroleum Group (Phase 1)	17.6	Late 2022
Gansu	Qingyang Ruihua Energy Co., Ltd.	6.5	August 2022
Chongqing	Sinopec Marketing Chongqing	11	November 2022

helium imports came from Qatar, 10.61% from the United States, and 3.96% from Australia. Compared to 2021, imports from Qatar and the U.S. increased, while those from Australia declined.

In recent years, influenced by both international and domestic exploration and development factors, helium import prices have fluctuated dramatically, as shown in Fig. 2.5. From 2019 to 2023, the price trends for liquid helium and high-purity helium imported by China were roughly the same. For instance, in 2019, the tight supply of liquid helium worsened, causing prices to soar to RMB 230/L in

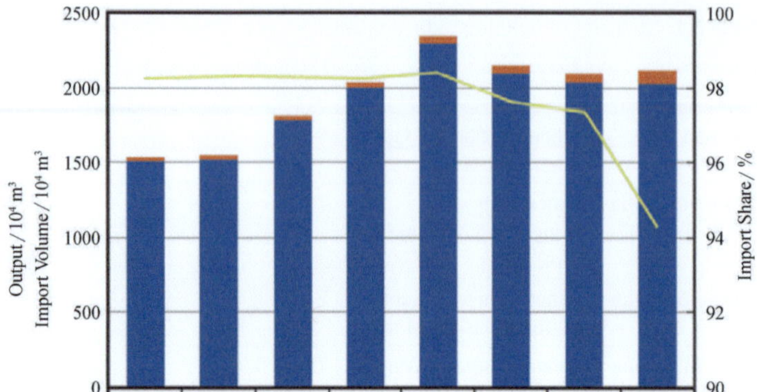

	2014	2015	2016	2017	2018	2019	2020	2021
Output / 10^4 m^3	24	25	28	32	35	50	53	130
Import Volume / 10^4 m^3	1513	1589	1774	2009	2311	2226	2076	2065
Import Share / %	98.44	98.45	98.45	98.43	98.51	97.80	97.46	94.07

Fig. 2.3 China's helium output and import volume trend

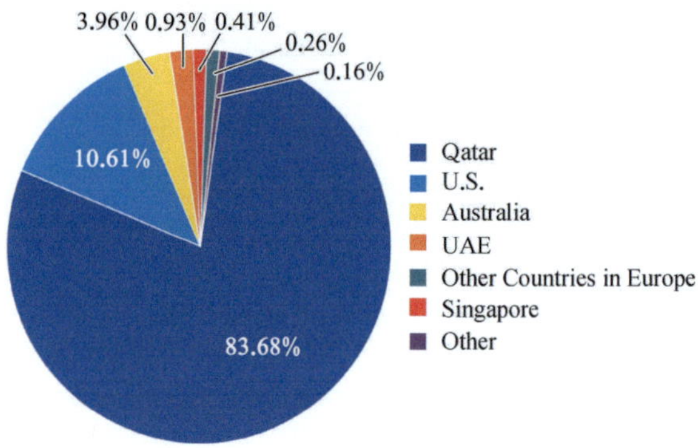

Fig. 2.4 China's helium imports in 2022 (by trade partner)

June. However, by August, the market cooled, and prices dropped to around RMB 210/L in September. The annual average price for liquid helium in 2019 was RMB 211.7/L, an 80.2% increase compared to 2018. In 2020, as demand from downstream industries weakened, liquid helium prices continued to decline, down to RMB 100/L by December. In 2021, with the recovery of downstream demand and limited supply in the short term, liquid helium prices rebounded, clocking up RMB 180/L by December, with an annual average price of RMB 126.25/L, an 11.66% decrease compared to 2020. In 2022, the average price of liquid helium rose to RMB 193.75/L, a 53.47% increase over 2021. However, in 2023, as demand fell short of expectations, the market declined, with the average price from January to October at RMB 162.5/L, down 13.33% compared to the same period last year.

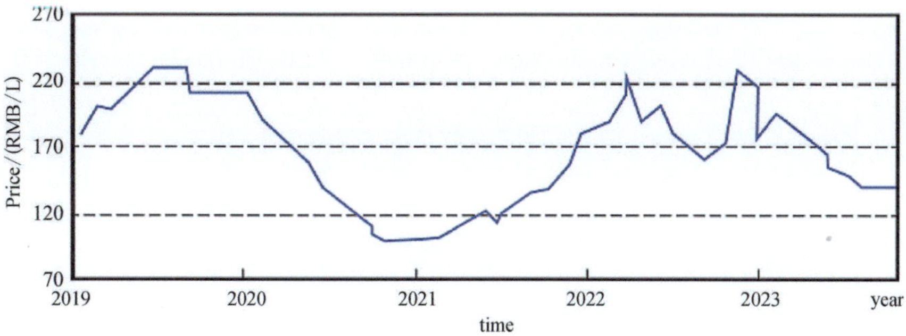

Fig. 2.5 Liquid helium market price trends from 2019 to 2023

2.3.3 Helium Consumption Sectors

From 2018 to 2022, China's helium consumption experienced a decrease followed by an increase. In 2020, the extensive shutdown of downstream industries led to weak demand for helium, with consumption falling to 2.292×10^7 m^3. However, in 2021, recovery in industries such as semiconductors and optical fibers boosted helium consumption to 2.358×10^7 m^3. Despite the impact of special circumstances in 2022 and varied performances in downstream industries, the international landscape heightened market tensions. The total helium consumption for the year was approximately 2.556×10^7 m^3.

The main destinations for helium imports in China are Jiangsu, Shanghai, Guangdong, and Zhejiang, which together account for 89% of total imports. The helium market is primarily concentrated in East China, Central China, North China, and South China, with these regions representing over 75% of the total consumption. In terms of consumption structure, the consumption of helium in China is concentrated in specific fields. In 2022, as shown in Fig. 2.6, helium demand was concentrated in NMR, welding shielding gas, and laboratory applications. NMR applications accounted for 28% of the total, a growth from 2021. Welding shielding gas represented 16%, showing a slight decline from the previous year. Laboratory applications made up 10%, with a slight increase compared to 2021. Leak detection, balloon gas, optical fibers, and electronic semiconductors accounted for 4%–9%, with significant growth in optical fiber demand, while semiconductor and leak detection gases saw slight declines. Additionally, helium applications in cryogenics and deep-sea diving remained stable.

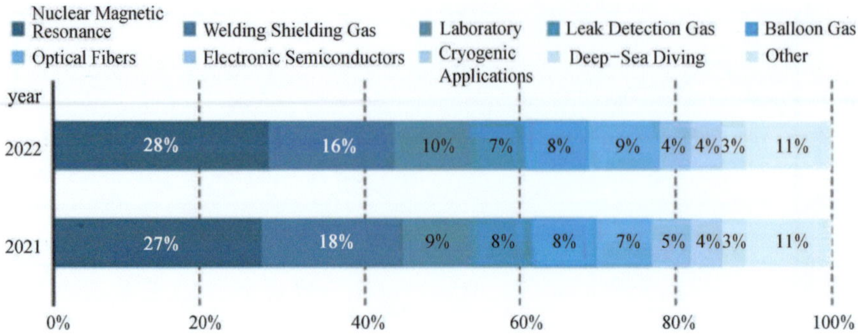

Fig. 2.6 Comparison of China's downstream helium demand structures in 2021–2022

2.4 Protection of Helium Resources

The United States was the first country to industrially extract helium and has long been the world's leading producer and supplier of helium. It is also the first country to recognize the importance of protecting helium resources, with a century-old tradition of legislating for helium resource protection.

Since World War I, the U.S. government has recognized the significance of helium, particularly with the development of long-range manned airships. Helium gradually replaced hydrogen as the preferred lifting gas for airships due to its safety and non-flammable properties. In 1925, to secure a stable supply of helium for national defense needs, the U.S. Congress established the Federal Helium Program and passed the *Helium Act of 1925*, which authorized the U.S. Bureau of Mines to build and operate large-scale helium extraction and purification plants.

With the sharp increase in helium demand during and after World War II, the U.S. Congress passed the *Helium Act Amendments of 1960*, encouraging private natural gas producers to extract helium and sell it to the government. In October 1996, the U.S. Senate and the Department of State enacted the *Helium Privatization Act of 1996*. This act directed the U.S. Department of the Interior and the National Academy of Sciences to make recommendations for revising the act and managing helium reserves if the helium industry, market, scientific research, technology applications, biomedicine, or national security were substantially subject to adverse impacts. The Federal Helium Program's newly defined responsibilities include the following four aspects:

(1) Operate and maintain helium storage facilities and pipeline systems.
(2) Supply crude helium to private companies through contracts.
(3) Assess the helium-containing natural gas resources of the U.S.
(4) Manage helium recovery and disposal on federal lands.

Additionally, the In-kind Crude Helium Sales Regulation (43 CFR part 3195), effective from December 23, 1998, requires that refiners selling helium to federal agencies and their contractors must purchase an equivalent amount of crude helium from the U.S. Bureau of Land Management (BLM) [21]. In 2013, the BLM enacted the *Helium Stewardship Act of 2013*. On February 16, 2018, then-U.S. President Donald Trump signed Executive Order No. 13817, directing the U.S. Department of the Interior and the Department of Defense to publish a list of 35 critical minerals vital to U.S. interests, which included helium.

In April 2019, the Australian federal government released *Australia's Critical Minerals Strategy 2019*, which enlisted 24 critical minerals important for defense, space, energy, and advanced manufacturing, with helium included. In November 2023, the European Union passed the *Critical Raw Materials Act*, setting out the latest list of critical and strategic raw materials. The act proposes that the EU bolster the security of critical raw material supplies, promote the innovation and application of substitute materials, and raise recycling standards for critical raw materials, with helium being listed as one of the key materials.

References

1. USGS. Mineral commodity summaries 2023. U.S. Geological Survey; 2023.
2. Broadhead RF. Helium in New Mexico; geologic distribution, resource demand, and exploration possibilities. New Mex Geol. 2005;27(4):93–101.
3. Halford DT, Karolyt R, Barry PH, et al. High helium reservoirs in the four corners area of the Colorado Plateau, USA. Chem Geol. 2022;596:120790.
4. Nuttall WJ, Clarke RH, Glowacki BA. The future of helium as a natural resource. Oxford, UK: Routledge; 2012.
5. Dai JX, Ni YY, Qin SF, et al. Geochemical characteristics of He and CO_2 from the Ordos (cratonic) and Bohaibay (rift) basins in China. Chem Geol. 2017;469:192–213.
6. Ni CH, Wu XQ, Liu QY, et al. Helium signatures of natural gas from the Dongpu Sag, Bohai Bay Basin, eastern China. Front Earth Sci. 2022;10:37–46.
7. Tang JR, Zhang YX, Zhou JL, et al. Analysis of global helium industry chain and China's strategy. Geol Bull China. 2023;42(01):1–13 (in Chinese)
8. Qin SF, Li JY, Wang JM, et al. Helium enrichment model of helium-rich gas reservoirs in petroliferous basins in China. Nat Gas Ind. 2022;42(7):125–134 (in Chinese).
9. He FQ, Wang FB, Wang J, et al. Helium distribution of Dongsheng gas field in Ordos Basin and discovery of a super large helium-rich gas field. Pet Geol Exp. 2022;44(1):1–10 (in Chinese).
10. Yu QX, Shi Z, Wang DG, et al. Analysis helium enrichment characteristics and reservoir forming conditions in Northwest Tarim Basin. Northwest Geol. 2013;46(4):215–222 (in Chinese).
11. Yu QX (2017) Oil and gas development should also consider helium resource utilization. China Energy News 14 (in Chinese).
12. Tao XW, Li JZ, Zhao LB, et al. Helium resources and the discovery of the first supergiant helium reserve in China: Hetianhe Gas Field. Earth Sci. 2019;44(3):1024–1041 (in Chinese).
13. Zhang YP, Li YH, Lu JC, et al. The Discovery and origin of helium-rich gas on the northern margin of the Qaidam Basin. Geol Bull China. 2016;35(2–3):364–371 (in Chinese).
14. Yang ZN, Li YH, Liu WJ, et al. Geological conditions of helium formation and resource prospect analysis in Quanjishan area, northern Qaidam Basin. Coal Geol China. 2018;30(6):64–70 (in Chinese).
15. Han W, Liu WJ, Li YH, et al. Characteristics of rare gas isotope and main controling factors of radon enrichment in the northern margin of Qaidam Basin. Nat Gas Geosci. 2020;31(3):385–392 (in Chinese).

16. Zhang XB, Zhou F, Cao ZY, et al. Finding of the Dongping economic helium gas field in the Qaidam Basin, and helium source and exploration prospect. Nat Gas Geosci. 2020;31(11):1585–1592 (in Chinese).
17. Li YH, Zhou JL, Zhang W, et al. Reservoir formation conditions and resource potential of helium in the Weihe Basin. Beijing: Geological Publishing House; 2018 (in Chinese).
18. Han YH, Luo HY, Xue YZ, et al. Genesis and helium enrichment mechanism of geothermal water-associated gas in Weihe Basin. Nat Gas Geosci. 2022;33(2):277–287 (in Chinese).
19. Li J, Wang XB, Xu ZS, et al. Helium resources accumulation regulations and their development prospects in China. Nat Gas Geosci. 2024;35(5):851–868 (in Chinese).
20. Li JF, He LL, Chai LH. Present situation and suggestion of helium extraction from natural gas. Chem Eng Oil & Gas. 2018;47(4):41–44 (in Chinese).
21. Hao XP. Suggestions on the legislative protection of helium resources in China. Chem Eng Oil & Gas. 2009;38(3):393–395 (in Chinese).

Chapter 3
Industrial Chain of Helium Extraction from Natural Gas

Liming Pu and Yunqiang Chen

Although trace amounts of helium can be produced through nuclear reactions, helium cannot be chemically synthesized or produced industrially. Instead, it can only be obtained at varying purity through physical purification. The concentration of helium in the air is merely 5.2 mL/m^3 (ppmV). Although helium can be industrially separated using non-condensable gases from large-scale air liquefaction and separation units, extracting helium from helium-containing natural gas is economically more feasible and remains the primary source of helium today. Helium content in helium-containing natural gas can reach 0.3%–8%. In some cases, even if the helium content in natural gas is less than 0.1%, helium can accumulate in the BOG from large-scale natural gas liquefaction plants, making helium recovery from BOG economically attractive.

Since the discovery of helium in natural gas by the United States in 1905, followed by the exploration of industrial helium extraction from 1917 to 1921, the helium extraction industry has evolved over a century. With its advantages in resources, scientific research, technologies, production and applications, the United States has forged a sound industrial chain of extracting helium from natural gas. In addition to supplying large quantities of helium products worldwide, it has also led the development of the helium extraction industry chain. Since the 1960s, Chinese scientists and engineers developed a path that suited China's national conditions, building its own industrial chain of helium extraction from natural gas.

This chapter reviews the history of helium extraction in China and other countries, summarizes key segments of the industrial chain, including helium production, storage, and transportation, and offers insights into the future development of China's helium extraction industry chain.

L. Pu (✉) · Y. Chen
China Petroleum Engineering & Construction Corporation Southwest Company, Chengdu, China
e-mail: puliming_sw@cnpc.com.cn

Y. Chen
e-mail: chenyunqiang@cnpc.com.cn

L. Pu (ed.), *Helium Engineering Technologies*,
https://doi.org/10.1007/978-981-96-4086-7_3

3.1 History of Extracting Helium from Natural Gas

Since the United States launched industrial helium extraction experiments during World War I, helium extraction from natural gas has seen more than a hundred years of development. China's helium extraction industry got off in the early 1960s. Currently, the helium extraction industry is primarily concentrated in a few countries with abundant helium-containing natural gas resources and high helium content, such as the United States, Qatar, and Russia.

3.1.1 History of Helium Extraction from Natural Gas in Other Countries

The extraction of helium from natural gas began in the early twentieth century. After more than a century of development, countries such as the United States and Germany have mastered helium extraction technologies.

On December 7, 1905, Hamilton Perkins Cady, a chemistry professor at the University of Kansas, first discovered helium in natural gas [1].

In 1915, the British launched a research project at the University of Toronto, and by 1918, they had built the world's first helium plant near Hamilton, proving the viability of helium separation technology. Through repeated experiments and numerous modifications, they finally separated helium from natural gas by mid-1918. The plant operated with helium purity clocking up 87%, with nitrogen as the primary residue.

In July 1917, the U.S. government launched a demonstration plant project for helium development. To maintain secrecy, the plants were referred to as "argon" plants during construction. In 1918, the U.S. established Argon Plant No. 1 and Argon Plant No. 2, experimental facilities north of Fort Worth, Texas, with helium purity of 92% and 70%, respectively. Shortly thereafter, the U.S. built Argon Plant No. 3 at the Petrolia Oil Field, but it could only produce helium with a purity of 20%. All three plants used natural gas from the Petrolia Oil Field, which had a helium content of about 1%. The commissioning of these three plants marked the United States' first steps toward large-scale helium production.

In 1921, the U.S. Navy established a large-scale commercial helium extraction plant near Fort Worth.

In 1925, the operation of the Fort Worth facility was transferred to the U.S. Bureau of Mines, and the same year, the *Helium Act* was established to ensure the supply of helium to the federal government. In March 1927, the *Amend an Act Authorizing Conservation of Helium Gas*, etc. was enacted. In April 1929, the Amarillo Helium Plant was built in the Cliffside gas field northwest of Amarillo, Texas, and the first shipment of approximately 5,663 m^3 of helium products was sent out on May 6th of that year [2].

In 1960, to conserve helium wasted during natural gas combustion, the U.S. Congress amended the *Helium Act*, allowing the government to sign long-term contracts to purchase crude helium from private companies for storage. In the same year, a helium pipeline was constructed from Bushton, Kansas, connecting several private helium recovery facilities to the Bush Dome storage reservoir. In 1963, private plants began producing helium in large quantities, allowing them to sell helium at prices lower than those of the federal government. The U.S. Bureau of Mines began storing privately owned crude helium in the Cliffside gas field, establishing it as the Federal Helium Reserve, located near Amarillo, Texas. In 1996, the U.S. passed the *Helium Privatization Act*, which allowed the federal government to halt production and sales of Grade A helium and sell off all crude helium in the Cliffside gas field. However, the 1996 *Helium Privatization Act* did not set a clear direction for the federal helium program after 2015, hence its amendment in 2013. The *Helium Stewardship Act of 2013* sought to privatize the Federal Helium Reserve through a competitive market process, thereby ensuring stability in the helium market. At that time, global helium production capacity was primarily concentrated in the United States, with U.S. helium plants listed in Table 3.1.

In the 1980s, the Soviet government decided to establish a helium production base in Orenburg to meet its helium needs. The helium content of the natural gas at the Orenburg condensate field was about 0.055%, with significant extraction potential.

Table 3.1 U.S. helium plants [1, 3]

Type (as of 1996)	Plant, location	$10^4 m^3/a$ Helium capacity/ (10^4 m^3/a)	Feed gas composition/% (mole fraction)		Commissioning year
			Helium	Nitrogen	
Government plants	Amarillo, Texas	170	1.80	25	1929
	Exell, Texas	679	0.90	10–18	1943
	Otis, Kansas	142	0.70–1.40	15	1943
	Shiprock, New Mexico	142	5.80	80	1944
	Keyes, Oklahoma	850	2.00	27	1959
	Kerr-McGee, Arizona	–	8.20	90	1962
Private plants (mainly pipeline gas)	Ulysses, Kansas	–	0.43	–	1962
	Liberal, Kansas	9000	0.30	–	1963
	Durnas, Kansas	11,000	0.66–0.71	–	1963
	Hansford, Texas	–	0.66–0.71	–	1963

The Orenburg helium plant consisted of three phases, with five helium units commissioned between 1979 and 1989, achieving a maximum helium production capacity of 9.20×10^6 m^3/a. To store the large quantities of helium extracted over the long term, the Soviet government established the Orenburg helium mixture salt cavern storage facility from 1979 to 1991.

In December 1976, Sonatrach Petroleum Corporation of Algeria planned to construct a potential helium project in Arzew, Algeria. After approximately 18 years of negotiation and construction, the Arzew helium plant was completed in 1995, and by the end of that year, it produced its first batch of liquid helium for export. From 1995 to 2009, the plant shipped over 1.42×10^8 m^3 of liquid helium, mainly sold to the EU market.

Since the twenty-first century, the rapid development of the liquefied natural gas (LNG) industry has significantly boosted international helium production capacity. Qatar's substantial LNG output, in particular, has long accounted for a quarter of the world's total, with its helium production capacity grows rapidly. The first phase of Qatar's helium plant was commissioned in September 2005, with a helium capacity of 1.864×10^7 m^3/a [4]. The second phase was commissioned in June 2013, with a helium capacity of 3.755×10^7 m^3/a. The third phase was commissioned in 2021, with a helium capacity of 1.200×10^7 m^3/a. These plants extract helium from BOG from upstream LNG plants.

In 2006, Algeria constructed a helium plant in Skikda with a helium capacity of 1960×10^4 m^3/a, producing helium from natural gas with a helium content of 4%–6%. The project was expanded in 2012.

In 2010, Australia commissioned the southern hemisphere's first helium production plant in Darwin [5], with a liquid helium output of 860 L/h, equivalent to approximately 4.3×10^6 m^3/a of helium.

In 2015, Gazprom began constructing the Amur plant in Russia, with a helium capacity of about 6×10^7 m^3/a. The first and second helium extraction units were commissioned in September 2021, and the third unit began trial production in August 2023.

In 2023, Russia commissioned the Irkutsk helium plant, with a helium capacity of 7.5×10^6 m^3/a. Currently, global helium production capacity remains relatively concentrated, with major helium plants worldwide listed in Table 3.2.

3.1.2 History of Extracting Helium from Natural Gas in China

In recent years, with the rapid development of high-tech industries, domestic demand for helium in China has witnessed a significant increase. After decades of development, China has initially mastered the complete set of technologies and supporting equipment for helium extraction from natural gas, which has been put into use.

Table 3.2 Major helium plants worldwide

No.	Country	Plant name	Helium capacity/(10^6 m^3/a)	Helium extraction process	Commissioning year
1	Poland	Poland helium plant	4	Membrane separation + PSA	1977
2	Algeria	Skikda helium plant	19.60	BOG	2006
3	Qatar	Qatar helium extraction plant (phase 1)	18.64	BOG	2005
4		Qatar helium extraction plant (phase 2)	37.55	BOG	2013
5		Qatar helium extraction plant (phase 3)	12	BOG	2021
6	Australia	Darwin helium extraction plant	4.30	BOG	2010
7	Canada	Mankota helium extraction plant	1.20	Membrane separation + PSA	2016
8	South Africa	Virginia natural gas project	0.71	BOG	–
9	Russia	Russia's amur gas processing plant	60	Ethane Recovery Co-Production	2021–2023

In the 1960s, the Sichuan Petroleum Administration Bureau, in cooperation with Sichuan Petroleum Survey & Design Institute under the former Ministry of Petroleum Industry (now China Petroleum Engineering & Construction Corporation Southwest Company), designed and built China's first helium extraction test unit I, using helium-rich natural gas from the Weiyuan Gas Field in the Sichuan Basin as the source. The plant had a design capacity of 5×10^4 m^3/d, with a helium output of 2×10^4 m^3/a. In 1989, based on this test unit, China National Petroleum Corporation (CNPC) built the Weiyuan Helium Extraction Test Unit II, with a design capacity of 1×10^5 m^3/d and a helium output of 4×10^4 m^3/a [6]. In 2012, Chengdu Natural Gas Chemical Works of PetroChina Southwest Oil & Gasfield Company built a new helium extraction plant in Dongxing Town, Rongxian County, Sichuan Province. Designed by China Petroleum Engineering Co., Ltd. Southwest Company (now China Petroleum Engineering & Construction Corporation Southwest Company), and it was the only operational helium extraction plant in China before 2020. The plant was designed to process 4×10^5 m^3/d of natural gas with a helium content of 0.18%, yielding an annual helium production capacity of about 2.1×10^5 m^3. The crude helium, with a purity of 90%–95%, was transported by cylinder trailers to the Chengdu Natural Gas Chemical Works for purification and refining into products with a purity of 99.999% [6].

Entering the twenty-first century, China's domestic demand for helium has risen significantly, leading to the successive launch of helium extraction projects utilizing BOG from LNG plants. In August 2020, the BOG helium extraction unit at the LNG plant of Inner Mongolia Xingsheng Natural Gas Co., Ltd. was put into operation, producing high-purity helium products. In October 2020, a BOG helium extraction project, jointly developed by Beijing Sinoscience Fullcryo Technology Co., Ltd. and Ningxia Shenran Zhongyuan Natural Gas Co., Ltd., was completed and put into operation in Yanchi County, Ningxia, with a liquid helium production capacity of 42.9 L/h.

3.2 Key Links in the Industrial Chain of Extracting Helium from Natural Gas

The helium industry chain, from production to application, mainly includes key steps such as helium resource exploration, helium production, helium storage, helium transportation, helium sales and users, as illustrated in Fig. 3.1. Helium is sourced from natural gas in helium-containing gas fields, BOG from natural gas liquefaction plants that use helium-containing natural gas as raw material, and purge gas from chemical plants using helium-containing natural gas for hydrogen production. The typical composition of helium sources in China is shown in Table 3.3.

Helium production can be divided into helium extraction from natural gas, helium extraction from BOG, helium extraction from purge gas and etc., depending on the gas source. According to the production stage, it falls into crude helium extraction, crude helium refining, and helium liquefaction. To be specific, crude helium extraction involves removing most of the methane, nitrogen, and other components from

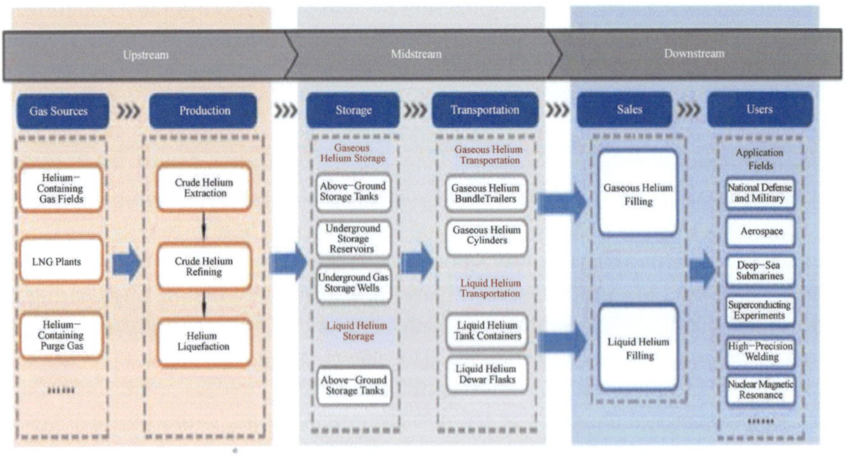

Fig. 3.1 Industrial chain of extracting helium from natural gas

Table 3.3 Typical composition of helium sources in China

No.	Component	Natural gas/% (mole fraction)	BOG/% (mole fraction)	Purge gas/% (mole fraction)
1	Methane	88.1880	81.7230	–
2	Ethane	2.2402	–	–
3	Propane	0.3500	–	–
4	Isobutane	0.0600	–	–
5	n-Butane	0.0700	–	–
6	Isopentane	0.0200	–	–
7	n-Pentane	0.0200	–	–
8	C_6^+	0.0400	–	–
9	Carbon dioxide	0.0700	–	–
10	Nitrogen	8.8008	14.8210	20–25
11	Helium	0.1340	3.2650	1.5–5
12	Hydrogen	0.0070	0.1397–0.1398	65–73
13	Oxygen	–	0.0510	–
14	Argon	–	–	1.5–5
15	Neon	–	0.0002–0.0003	0.0001–0.0002

helium-containing natural gas to produce crude helium. Crude helium refining is the process of purifying crude helium into helium products, while helium liquefaction involves condensing helium from a gaseous state to a liquid state. Helium storage can be categorized into gaseous helium storage and liquid helium storage, depending on the phase of helium being stored. Helium transportation refers to the process of transporting gaseous or liquid helium products to helium users. Based on transportation distance, helium products are transported to distribution points in the form of gaseous or liquid helium, and then delivered to users.

This section compares the technical status at home and abroad, introducing key aspects of the helium extraction industry chain, including helium production, helium storage, and product transportation, as illustrated in Fig. 3.2. Helium sources and users have been discussed in Sect. 2.1 (Distribution of Helium Resources) and Sect. 2.3 (China's Helium Market) and, therefore, will not be detailed here.

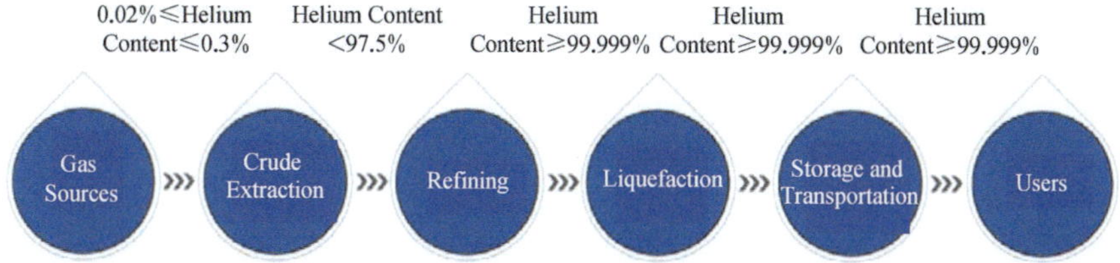

Fig. 3.2 Technical roadmap for helium extraction from natural gas

3.2.1 Crude Helium Extraction

Crude helium extraction refers to the process of concentrating the helium content in natural gas to a level below 97.5%. Typically, crude helium extraction is accomplished using cryogenic technology, membrane separation technology, or a combination of both. Currently, the focus of crude helium extraction has shifted towards multi-product co-production technologies, diversifying the range of products to include helium, LNG, ethane, and other sources of raw materials.

1. Cryogenic Technology

Cryogenic technology utilizes the difference in boiling points of the various components in natural gas and the fact that helium has a much lower liquefaction temperature compared to other components. Under certain temperature and pressure conditions, the other components are condensed to separate and concentrate the helium. This technology is one of the widely adopted approaches to extract crude helium from natural gas globally [7] and is relatively mature. It can generally concentrate helium in helium-containing natural gas to 50%–80%. Cryogenic technology supports a wide range of raw materials and exhibits mature processes, yet it requires significant investment, making it suitable for large-scale helium extraction plants.

When using cryogenic technology, carbon dioxide, hydrogen sulfide, and water must be removed from the helium-containing natural gas to prevent freezing and blocking of equipment and pipelines under low-temperature conditions. Cryogenic technology can be classified into throttling flash evaporation processes and cryogenic rectification processes based on the means of separation. The latter can further be categorized into single-column rectification, dual-column rectification, and three-column rectification, depending on the number of columns [8, 9]. For example, the helium extraction plant in Rongxian County, China, uses a dual-column cryogenic rectification process to produce helium products.

2. Membrane Separation Technology

Membrane separation technology is based on the high permeability of helium. It utilizes the differences in permeability of the various components in natural gas during the dissolution-diffusion process through a membrane to achieve separation and purification under the driving force of pressure differentials across the membrane. Typically, single-stage membrane separation can concentrate helium in helium-containing natural gas by 8–12 times. Compared to cryogenic technology, membrane separation technology does not involve phase change, making the process simpler and more suitable for small-scale helium extraction plants.

Currently, hollow fiber membranes or wound membranes made of polyimide are commonly used for helium extraction from natural gas. Countries such as the United States, Japan, Germany, and France have rolled out mature commercial membranes. China has also developed organic commercial membranes made of polyimide material.

3. Combined Membrane Separation and Cryogenic Technology

Due to variations in the composition and pressure of helium-containing natural gas, product demand, and site conditions, a combination of cryogenic technology and membrane separation technology can be employed to extract crude helium. This approach lowers investment and energy consumption while improving econonic benefits by leveraging the strengths of both technologies.

4. Multi-product Co-production Technology

During the cryogenic crude helium extraction process, hydrocarbons such as methane, ethane, and propane, as well as nitrogen, are condensed. To improve the economic feasibility of helium extraction, the process can be integrated with natural gas liquefaction and NGL recovery processes to produce LNG, ethane, liquefied petroleum gas (LPG), and stabilized light hydrocarbons. Therefore, based on whether co-production of value-added products is involved, cryogenic technology can be divided into single cryogenic crude helium extraction processes, NGL recovery and helium extraction integrated processes, natural gas liquefaction and helium extraction integrated processes, and NGL recovery combined with cryogenic rectification for co-production of LNG. Russia's Amur plant, for instance, adopts the NGL and helium extraction integrated process, co-producing ethane, LPG, and condensate products while extracting helium.

5. Multiple Gas Source Technology

During natural gas liquefaction, all hydrocarbons such as methane, ethane, and propane are condensed to produce LNG. Since the condensation point of helium is much lower, helium will be enriched in the BOG generated during natural gas liquefaction, where the helium content can reach 0.5%–3%. Qatar's Ras Laffan helium plant, for example, extracts helium from the BOG produced at LNG plants.

In synthetic ammonia units using helium-containing natural gas as raw material, a certain amount of purge gas is generated, with helium content reaching 0.2%–3%, which has extraction value. China has conducted helium recovery from purge gas in natural gas chemical processes and has established industrial facilities using synthetic ammonia purge gas to extract helium, producing helium products with a purity of 99.999%.

China's overall technical capabilities for crude helium extraction have reached an international advanced level. However, there is still a gap in the commercialization of helium separation membranes. A comparison of crude helium extraction technologies between China and other countries is provided in Table 3.4.

Table 3.4 Comparison of crude helium extraction technologies between China and other countries

Key technology	Status in other countries	Status in China
Cryogenic technology	Germany and the United States etc., have mature cryogenic helium extraction equipment with numerous engineering applications. Key parameters: (1) Cryogenic helium extraction processes include single-column rectification, dual-column rectification, etc.; (2) Engineering application example: the Amarillo helium plant in the United States with a helium production capacity of 1.70×10^6 m^3/a	There are engineering applications in China, and technology and equipment are relatively mature. Key parameters: (1) Cryogenic helium extraction processes include single-column rectification, dual-column rectification, etc.; (2) Engineering application example: a helium extraction plant with a helium production capacity of over 1.00×10^6 m^3/a
Membrane separation technology	The United States, Japan, Germany, and France have various mature commercial helium separation membranes. Key parameters: (1) Polyimide hollow fiber membranes or wound membranes; (2) Helium recovery rate: 80%–95%; (3) Operating temperature $\leq$ 90 °C; (4) Helium concentration ratio: 8–12; (5) Engineering application example: the Mankota helium extraction plant in Canada, with a helium production capacity of 1.2 $\times 10^6$ m^3/a	China has developed commercial helium separation membranes. Key parameters: (1) Polyimide hollow fiber membranes; (2) Helium recovery rate: 80%–90%; (3) Operating temperature $\leq$ 90 °C; (4) Helium concentration ratio: 7–10; (5) Engineering application example: a helium extraction plant which uses imported membranes with a helium production capacity of 7×10^4 m^3/a

(continued)

Table 3.4 (continued)

Key technology	Status in other countries	Status in China
Multi-product co-production technology	Germany, the United States, and other countries have mature equipment for natural gas NGL recovery combined with helium extraction, as well as engineering achievements. Key parameters: (1) Helium extraction process: ethane recovery combined with helium extraction; (2) Engineering application example: the Amur plant in Russia, with a helium production capacity of 6×10^7 m^3/a	There are engineering achievements in China regarding natural gas liquefaction and natural gas NGL recovery combined with helium extraction. Key parameters: (1) Helium extraction process: natural gas liquefaction combined with helium extraction; (2) Engineering application example: a helium extraction plant with a helium production capacity of over 1×10^6 m^3/a
Multiple gas source crude helium extraction technology	Germany, the United States, and other countries have mature LNG BOG helium extraction equipment and engineering applications. Key parameters: (1) Helium extraction source: BOG; (2) Helium extraction process: membrane separation + PSA and cryogenic rectification; (3) Engineering application example: Qatar helium extraction plant, with a helium production capacity of 6.819×10^7 m^3/a	Domestic BOG and purge gas helium extraction technology and equipment have seen engineering applications. Key parameters of BOG helium extraction: (1) Helium extraction process: membrane separation + cryogenic rectification; (2) Engineering application example: a BOG helium extraction project with a helium production capacity of 7.5×10^5 m^3/a Key parameters of purge gas helium extraction: (1) Helium extraction process: cryogenic helium extraction process; (2) Engineering application example: the purge gas helium extraction facility with a helium production capacity of 1.2×10^5 m^3/a

3.2.2 Crude Helium Refining

Crude helium refining involves further purifying crude helium to produce industrial helium, pure helium, high-purity helium, and ultra-pure helium products. Due to variations in the composition of helium sources, different approaches to helium extraction are employed, leading to varying levels of impurities in the extracted crude helium.

For example, crude helium extracted from a particular helium-containing natural gas using cryogenic technology typically contains impurities such as nitrogen, methane, and hydrogen. On the other hand, crude helium extracted using membrane separation technology generally contains impurities such as methane, ethane, propane, nitrogen, hydrogen, and carbon dioxide etc.

Since hydrogen has a similar liquefaction temperature and molecular size to helium, cryogenic technology, membrane separation technology, and PSA technology cannot separate helium from hydrogen. In crude helium refining, the reducing properties of hydrogen are often leveraged. Under this premise, oxygen is injected into the crude helium, triggering a catalytic reaction that produces water and removes the hydrogen. To secure complete removal of hydrogen, multi-stage catalytic oxidation reactors are usually set up, and excess oxygen is injected. As a result, after hydrogen removal, water and oxygen impurities are introduced into the crude helium once again.

Water, carbon dioxide, hydrocarbons, and nitrogen in crude helium can be removed through PSA. PSA involves selecting different adsorbent materials to eliminate various components of the crude helium. Silica gel or alumina bed layers, for example, are used for water removal, activated carbon bed layers for carbon dioxide removal, and zeolite bed layers for methane and nitrogen removal. Once the adsorbent is saturated with the impurity gases, the adsorbed gases can be desorbed by lowering the pressure, back-blowing, or applying vacuum, thereby regenerating the adsorbent's capacity. Depending on the adsorption operating pressure, the adsorbed components, and the adsorbent used, temperature swing adsorption (TSA) technology may also be employed.

If the helium-containing natural gas contains neon, cryogenic technology, membrane separation technology, and PSA technology cannot completely remove neon, which will accumulate as helium concentration increases. In industrial applications, neon is typically removed by subjecting high-pressure crude helium to cryogenic adsorption at liquid nitrogen temperatures. Cryogenic adsorption technology can also remove impurities such as argon and nitrogen from crude helium, resulting in high-purity helium products with a purity of no less than 99.999%. Depending on the desorption method used, low-temperature PSA technology may also be applied. China's technical capabilities in crude helium refining have now reached international advanced levels. A comparison between China's crude helium refining technology and that of other countries is presented in Table 3.5.

3.2.3 Helium Liquefaction

Helium liquefaction is the process of condensing gaseous helium into a liquid state at low temperatures. Helium has an extremely low boiling point, making it the most difficult gas to liquefy in nature. In 1908, physicist Heike Kamerlingh Onnes at the Physics Laboratory of Leiden University in the Netherlands liquefied helium for the first time, opening a new era in cryogenic engineering and low-temperature physics.

Table 3.5 Comparison of crude helium refining technologies between china and other countries

Key technology	Status in other countries	Status in China
Catalytic dehydrogenation technology	The U.S. and Switzerland possess mature technology and engineering experience. Key parameters: (1) Catalytic oxidation dehydrogenation process; (2) Post-treatment H_2 content ≤ 0.1 mL/m^3 (ppmV)	China has mastered and applied the technology and equipment. Key parameters: (1) Catalytic oxidation dehydrogenation process; (2) Post-treatment H_2 content ≤ 0.1 mL/m^3 (ppmV)
Pressure swing adsorption technology	The U.S., Germany, and France have mature technology and engineering experience. Key parameters: Post-treatment helium content 99%–99.99%	China has mastered and applied the technology and equipment. Key parameters: Post-treatment helium content 99%–99.99%
Low-temperature adsorption technology	The U.S., Germany, and France have mature technology and engineering experience. Key parameters: Post-treatment helium content $\geq$ 99.999%	China has mastered and applied the technology and equipment. Key parameters: Post-treatment helium content $\geq$ 99.999%

Onnes utilized a high-pressure throttling refrigeration cycle, where high-pressure helium was sequentially passed through a liquid nitrogen heat exchanger, a liquid hydrogen heat exchanger, and further cooled by a helium BOG heat exchanger. The gas then expanded isenthalpically through a throttle valve, and after gas–liquid separation, liquid helium was produced, with a production rate of merely 250 mL/h. Today, the industry mainly uses helium cycle liquefaction technology with liquid nitrogen precooling and multi-stage expansion refrigeration, which is far more efficient than high-pressure throttling refrigeration cycles, and helium liquefaction capacity has been significantly boosted.

Globally, the main producers of helium liquefiers are Air Liquide in France and Linde in Germany. Air Liquide offers standard helium liquefiers with liquid helium production capacities ranging from 30 L/h to 330 L/h, and can also customize helium liquefiers with production capacities ranging from 330 L/h to 3100 L/h based on demand. Linde's standard helium liquefier series, including the L70, L140, and L280, have production capacities of 40–70 L/h, 90–140 L/h, and 200–290 L/h, respectively. Custom helium liquefiers with capacities ranging from 350 L/h to 3500 L/h can also be provided. In China, the main helium liquefier manufacturer is the Technical Institute of Physics and Chemistry of the CAS (Chinese Academy of Science), which offers standard helium liquefiers with production capacities ranging from 40 L/h to 300 L/h. A comparison of China's helium liquefaction technology/equipment with that of other countries is shown in Table 3.6.

Table 3.6 Comparison of helium liquefaction technology/equipment between China and other countries

Key technology/equipment	Status in other countries	Status in China
Helium liquefiers	Germany and France possess the capability and engineering experience to manufacture helium liquefiers with capacities of 800 L/h and above. Key parameters: (1) Helium liquefaction process with liquid nitrogen precooling and two-stage expansion refrigeration; (2) Single liquefier production capacity of 30–3500 L/h	China has developed 300 L/h helium liquefiers, which have seen engineering applications. Key parameters: (1) Helium liquefaction process with liquid nitrogen precooling and three-stage expansion refrigeration; (2) Single liquefier production capacity of 300 L/h

3.2.4 *Helium Storage and Transport*

Helium storage and transport can be categorized by the phase of helium into gaseous way and liquid way. Based on the pressure of the storage and transportation equipment, it can be classified into high-pressure and low-pressure storage and transport. Additionally, storage methods can be divided into above-ground and underground storage. Gaseous helium is typically stored in high-pressure containers, which include fixed and mobile pressure vessels such as storage tanks, tube bundles, and gas cylinders. Low-pressure storage mainly involves gas holders, which are commonly used for recovering residual gas. China's technical capabilities in gaseous helium storage and transport have reached advanced international levels. A comparison of gaseous helium storage and transportation methods between China and other countries is shown in Table 3.7.

Since 1 m^3 of liquid helium can evaporate into 750.96 m^3 of gaseous helium (at 20 °C, 101.325 kPa), liquid helium products are more suitable for storage and long-distance transportation. Liquid helium storage and transportation equipment includes liquid helium tank containers and liquid helium Dewar. Currently, the countries capable of producing large liquid helium tank containers are the United States and Germany. China has manufactured a mobile liquid helium tank container with a volume of 40 m^3 and a fixed one with a volume of 10 m^3, while also developing a fixed liquid helium tank container with a volume of 120 m^3. A comparison of liquid helium storage and transportation technologies between China and other countries is shown in Table 3.8.

Table 3.7 Comparison of gaseous helium storage and transportation methods between China and other countries

Key technology/equipment	Status in other countries	Status in China
Underground helium storage	The United States and Russia have established large-scale national underground helium storage facilities. Key parameters: (1) Helium concentration: 20%–60%; (2) Storage pressure: 10–25 MPa	China has initiated research on helium storage facilities. Key parameters: (1) Helium concentration: about 10%; (2) Storage pressure: about 10 MPa
Above-ground gaseous helium storage and transport	Western countries such as the United States, Germany, and France have mastered and applied high-pressure helium storage tank and cylinder technology and equipment. Key parameters: (1) Storage method: high-pressure tanks or cylinders; (2) Operating volume: 4–28 m^3; (3) Operating pressure: 20–35 MPa	China has mastered and applied high-pressure helium storage tank and cylinder technology and equipment. Key parameters: (1) Storage method: high-pressure tanks or cylinders; (2) Operating volume: 4–28 m^3; (3) Operating pressure: 20–35 MPa

3.3 Helium-Related Standards

In the helium industry, standards, specifications and technical manuals serve as an important guide throughout the chain. The U.S. Defense Logistics Agency (DLA) (Richmond Aviation) compiled the *Helium Plants and Storage* manual (MIL-HDBK-1024/2) in 1987, which primarily sets out requirements for helium plants and storage equipment. In China, several standards have been released, including those related to helium testing, leak detection, helium products, and associated equipment. During the preparation of this book, the industry standard *Design Specifications for Helium Extraction from Natural Gas* (*SYT 7780-2024*), edited by China Petroleum Engineering & Construction Corporation Southwest Company, was released on September 24, 2024, and will come into effect on March 24, 2025.

Currently, standards for the helium industry are listed in Table 3.9.

Table 3.8 Comparison of liquid helium storage and transportation technologies between China and other countries

Key technology/equipment	Status in other countries	Status in China
Fixed liquid helium tank containers	Germany and the United States have the capability and engineering experience to manufacture fixed liquid helium tank containers. Key parameters: (1) Operating temperature: –269 °C; (2) Operating volume: 60–120 m^3	China has developed fixed liquid helium tank containers with volumes of 10 m^3 and 120 m^3. Key parameters: (1) Operating temperature: –269 °C; (2) Operating volume: 10 m^3, 120 m^3
Mobile liquid helium tank containers	Germany and the United States have the capability and engineering experience to manufacture mobile liquid helium tank containers. Key parameters: (1) Operating temperature: –269 °C; (2) Operating volume: 40 m^3; (3) Holding time: 40–45 days; (4) Transportation method: automotive transport	China has developed mobile liquid helium tank containers with a volume of 40 m^3. Key parameters: (1) Operating temperature: –269 °C; (2) Operating volume: 40 m^3; (3) Holding time: 15–20 days; (4) Transportation method: automotive transport

Table 3.9 Relevant standards for the helium industry

No.	Standard name	Standard no. (source)
1	Helium plants and storage	MIL-HDBK-1024/2
2	Helium leak detector	AIR FORCE A-A-59870-2010
3	Leak detector, helium	A A 59355 Notice 1-Validation: 2012
4	Precision cleaning and testing of shipboard oxygen, helium, helium–oxygen, and nitrogen systems	MIL MIL-STD-1330D
5	Valve, oxygen, helium and helium oxygen mixture, high pressure for gas services	MIL MIL-V-24439A Amendment 1

(continued)

Table 3.9 (continued)

No.	Standard name	Standard no. (source)
6	Semitrailer, compressed gas cylinder, helium commercial	MIL MIL-S 62127 A
7	Valve, ball, naval shipboard for air, nitrogen, helium or hydraulic service (size 2 1/2 in. IPS and below)	MIL MIL-V-22687C
8	Valve, oxygen, helium and helium oxygen mixture, high pressure for gas services	NPFC-QPL 24439
9	Missile components, liquid oxygen, liquid nitrogen, gaseous oxygen, gaseous nitrogen, instrument air, helium and fuel handling systems; cleaning and packaging for delivery	NPFC-MIL-M-9950
10	Precision cleaning and testing of shipboard oxygen, helium, helium–oxygen, nitrogen and hydrogen systems	PPFC-MIL STD 1330
11	Valve, cylinder, gas; helium and nitrogen (inert-oil tolerant), Outlet 590 (pressures through 3000 Psig (20,680 kPa) at 120 °F (48.9 °C)	DLA-MIL-DTL-2/26 D
12	Specification for design of helium extraction from natural gas	SYT 7780-2024
13	Supervision regulation on safety technology for transportable pressure vessel	TSGR 0005-2011
14	Large-volume seamless steel gas cylinders	GB/T 33145-2016
15	Tube trailer	NB/T 10354-2019
16	Tube bundle container	NB/T 10355-2019
17	Oil-free reciprocating helium compressors	JB/T 13630-2019
18	Pure helium, high pure helium and ultra pure helium	GB/T 4844-2011
19	Gas for electronic industry—helium	GB/T 16943-2009
20	Industrial helium	GB/T 28123-2011
21	Mixed gas for welding carbon dioxide, oxygen/argon, helium/argon and oxygen/argon	HG/T 4984 4986 2016
22	Mixed gas for welding—helium/argon	HG/T 4985-2016
23	Mixed gas for laser welding—carbon dioxide, nitrogen/helium	HG/T 5898-2022

(continued)

Table 3.9 (continued)

No.	Standard name	Standard no. (source)
24	Non-destructive testing—test methods for helium leak testing	GB/T 15823-2009
25	Natural gas—determination of composition with defined uncertainty by gas chromatography	GB/T 27894-2011
26	Gas analysis—gas chromatograph with helium ionization detector	GB/T 28726-2012
27	Measurement of helium porosity and pulse decay permeability of shale	GB/T 34533-2017
28	Vacuum technology—vacuum leak testing method by helium mass spectrometer	GB/T 36176-2018
29	Calibration specification for vacuum helium leaks	JJF 1833-2020

References

1. Grynia E, Griffin PJ. Helium in natural gas-occurrence and production. J Nat Gas Eng. 2016;1(2):163–215.
2. Tade MD. Helium storage in cliffside field. J Petrol Technol. 1967;19(7):885–888.
3. Yakutseni VP. World helium resources and the perspectives of helium industry development. Нефтегазовая геология Теория и практика. 2014;9(1):1–22.
4. Daly JW. Helium recovery from LNG. In: International petroleum technology conference. IPTC; 2005.
5. Lindemann U, Boeck S, Blum L, et al. Turnkey helium purification and liquefaction plant for Darwin, Australia. AIP Conf Proc. 1218(1):271–274.
6. Li JF, He LL, Chai LH. Present situation and suggestion of helium extraction from natural gas. Chem Eng Oil Gas. 2018;47(4):41–44 (in Chinese).
7. Hamedi H, Karimi A, Gunderson T, et al. Optimization of helium extraction processes integrated with nitrogen removal units: a comparative study. Comput Chem Eng. 2019;121:354–366.
8. Rufford TE, Chan KI, Huang SH, et al. A review of conventional and emerging process technologies for the recovery of helium from natural gas. Adsorpt Sci Technol. 2014;32(1):49–72.
9. Elvers B, Hawkins S, Schulz G. Ullmann's encyclopedia of industrial. Hoboken, NJ: Verlag Chemie; 1991.

Chapter 4
Crude Helium Extraction Technologies

Ke Wang, Yingjie Xiong, and Liming Pu

Crude helium extraction serves as a pivotal link in the industrial chain of helium extraction from natural gas. This process requires the removal of most methane, nitrogen, and other components from helium-containing natural gas to produce crude helium with a concentration no higher than 97.5%. Typically, crude helium extraction is accomplished using cryogenic technology, membrane separation technology, or a combination of both. Currently, the focus is on multi-product co-production technologies, diversifying the range of products to include helium, LNG, ethane, and other sources of raw materials. Pressure Swing Adsorption (PSA) is also a widely applied method in industrial crude helium extraction. However, it is typically employed in combination with other technologies. PSA technology will be discussed in detail in Chap. 5 on crude helium refining technologies.

This chapter mainly introduces various crude helium extraction technologies, including cryogenic technology, membrane separation technology, combined membrane separation and cryogenic technology, multi-product co-production technology, and multiple gas source technology (Fig. 4.1).

K. Wang · Y. Xiong · L. Pu (✉)
China Petroleum Engineering & Construction Corporation Southwest Company, Chengdu, China
e-mail: puliming_sw@cnpc.com.cn

K. Wang
e-mail: wangke_sw@cnpc.com.cn

Y. Xiong
e-mail: xiongyingjie_sw@cnpc.com.cn

L. Pu (ed.), *Helium Engineering Technologies*,
https://doi.org/10.1007/978-981-96-4086-7_4

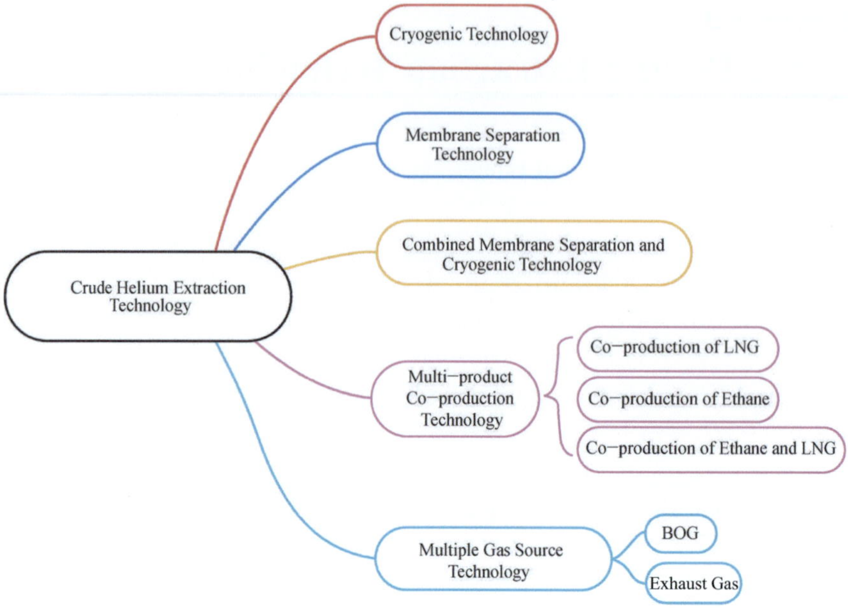

Fig. 4.1 Schematic diagram of crude helium extraction technologies

4.1 Cryogenic Technology

Cryogenic technology, also referred to as deep freezing or ultra-low temperature technology, typically pertains to scientific and technological applications involving temperatures below − 153 °C (120 K). The principle of cryogenic helium extraction technology lies in exploiting the differences in boiling points among the components in natural gas. Under low-temperature conditions, components such as methane and nitrogen are condensed and separated, thereby increasing the helium content in the gas phase. The boiling points of major components in natural gas at atmospheric pressure are shown in Fig. 4.2, where the boiling point of methane is − 161.48 °C, ethane − 88.58 °C, propane − 42.11 °C, carbon dioxide − 78.46 °C, nitrogen − 195.80 °C, hydrogen − 252.78 °C, and helium − 268.93 °C. Cryogenic technology is currently the most mature and widely applied method for crude helium extraction. It is suitable for large-scale helium extraction from natural gas and typically achieves a helium recovery rate of over 95%, offering high reliability and stability. Cryogenic technology primarily encompasses flash evaporation and distillation processes, and can also be implemented as a combination of both.

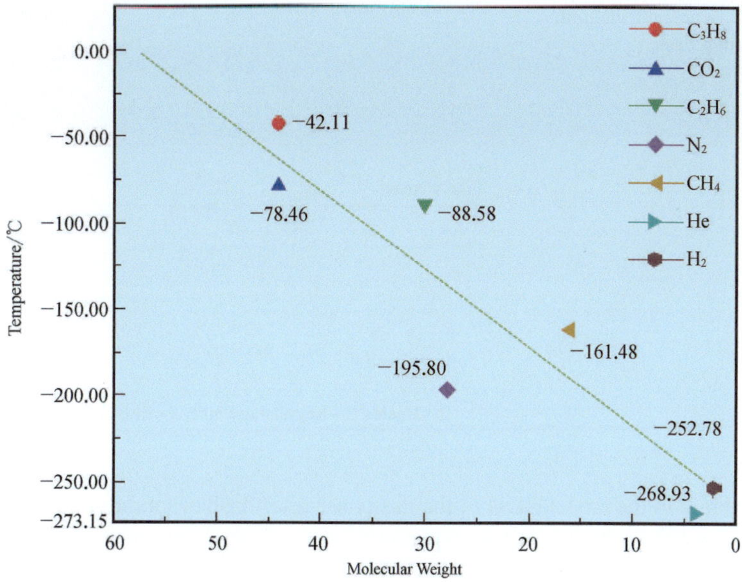

Fig. 4.2 Boiling point curves of major components in natural gas at atmospheric pressure

4.1.1 Flash Evaporation for Helium Extraction

Flash evaporation involves reducing pressure to cause the fluid to boil, thereby generating both gas and liquid phases and establishing a new equilibrium at a different pressure level. Since the boiling point of a substance increases with pressure, the process allows high-pressure, high-temperature fluid to lower its boiling point through depressurization. When the fluid's temperature is higher than the boiling point at the reduced pressure, it quickly boils and vaporizes in the flash tank, where phase separation occurs. The equipment responsible for vaporizing the fluid is not the flash tank itself but a throttle valve. The flash tank provides space for the rapid vaporization and separation of the gas and liquid phases. However, flash evaporation comes at a cost, primarily in terms of the loss of pressure energy. The schematic diagram of throttling flash evaporation is illustrated in Fig. 4.3.

For a mixture with overall composition z_1, z_2, …, z_N, when the system's temperature T and pressure p fall between the bubble point and dew point, two phases are automatically generated in gas–liquid equilibrium. The gas phase composition is y_1, y_2, …, y_N, and the liquid phase composition is x_1, x_2, …, x_N. In practice, flash evaporation is a single-stage equilibrium separation process. If the pressure of a liquid mixture, which is above the bubble point pressure, is lowered to a level between the bubble point and dew point pressures, it undergoes partial vaporization, resulting in flash evaporation. The flash evaporation process aligns with the principles of mass conservation and thermodynamic phase equilibrium, which include the overall material balance, individual component balance, and gas–liquid balance. If

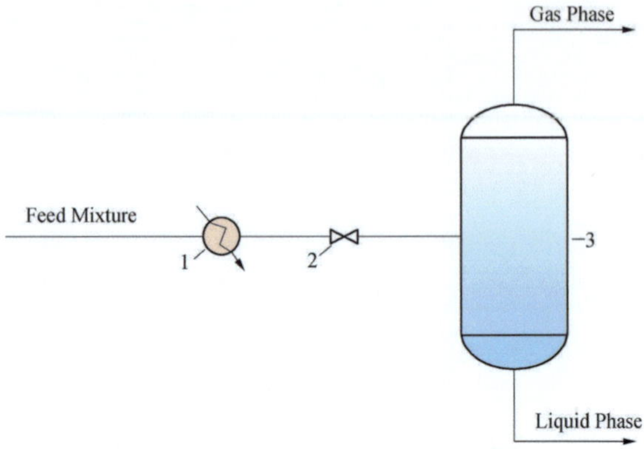

Fig. 4.3 Schematic diagram of throttling flash evaporation

1 - Heater; 2 - Regulating Valve; 3 - Rectification Column

the feed rate to the flash tank is F, the gas phase after flash evaporation is V, and the liquid phase is L, the calculation equations for the flash evaporation process are as shown in (4.1)–(4.6).

In the throttling flash evaporation process, the material balance equation reflects the mass relationship between the inlet mixture and the outlet liquid–gas phases:

$$F = L + V \tag{4.1}$$

$$F \cdot z_i = V \cdot y_i + L \cdot x_i \tag{4.2}$$

$$y_i = K \cdot x_i \tag{4.3}$$

where K is the gas–liquid equilibrium ratio or phase equilibrium ratio, representing the ratio of the mole fractions of a component in the gas phase to the liquid phase at equilibrium. Since the compositions of the gas and liquid phases y_i or x_i are not entirely independent, they must satisfy

$$\Sigma y_i = 1 \tag{4.4}$$

or

$$\Sigma x_i = 1 \tag{4.5}$$

Heat balance for the flash evaporation process is given by:

$$F \cdot H_{\mathrm{F}} + Q = V \cdot H_V + L \cdot H_{\mathrm{L}} \tag{4.6}$$

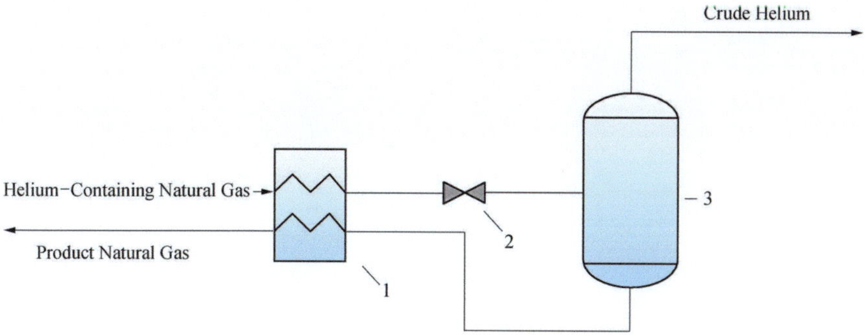

1 - Heat Exchanger; 2 - Regulating Valve; 3 - Separator

Fig. 4.4 Flow chart of single-stage flash evaporation

where H_F, H_V, and H_L represent the molar enthalpies (in joules or simply J) of the feed mixture, gas phase, and liquid phase, respectively, and Q is the heat in the process (in J). Flash evaporation for helium extraction is characterized by its low cost, high separation efficiency, low recovery rate, and low helium concentration in the crude helium. Flash evaporation can be classified into single-stage flash and multi-stage flash according to the number of flash stages. The process flow of single-stage flash evaporation for helium extraction is shown in Fig. 4.4. In this process, high-pressure helium-containing natural gas is subjected to heat exchange with refluxing low-pressure liquid hydrocarbons. The majority of nitrogen, methane, and other alkanes condense and undergo a single throttling flash evaporation. After gas–liquid separation in the separator, the liquid phase is refluxed and, after heat exchange with the high-pressure helium-containing natural gas, is exported as product gas. Due to the significantly lower boiling point of helium compared to nitrogen, methane, and other alkanes, helium in the gas phase is concentrated.

In the multi-stage flash evaporation process [1], the gas and liquid phases from the primary throttling flash evaporation, based on the single-stage flash evaporation process, undergo further condensation and additional stages of throttling flash evaporation to increase the helium content in the gas phase. The process flow of multi-stage flash evaporation for helium extraction is shown in Fig. 4.5. The pretreated helium-containing natural gas undergoes condensation and primary throttling before being introduced into the first separator to separate gas and liquid phases. The liquid phase from the first separator undergoes throttling and passes into the second separator for additional gas–liquid separation. Subsequently, the liquid phase from the second separator undergoes further throttling and enters the third separator to achieve gas–liquid separation. Upon throttling, the liquid phase from the third separator undergoes reheating in a heat exchanger before proceeding to the next processing step. The gas phases from the first, second, and third separators, after throttling, are sent as crude helium to the next process.

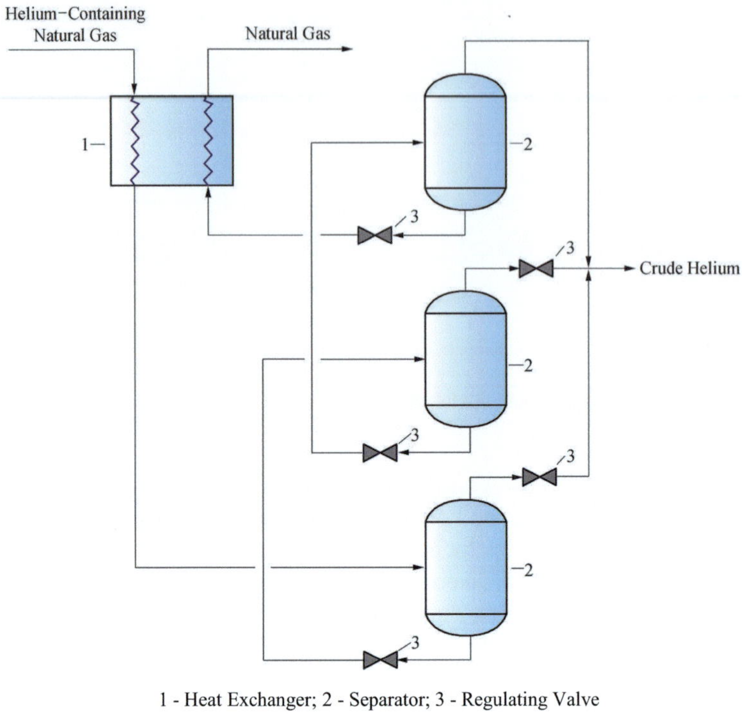

Fig. 4.5 Flow chart of multi-stage flash evaporation for helium extraction

4.1.2 Rectification for Helium Extraction

Rectification for helium extraction refers to the use of a rectification column at low temperatures to separate helium, nitrogen, hydrocarbons, and other components through repeated vaporization and condensation cycles under certain pressures. When helium-containing liquid hydrocarbons enter the rectification column, they are heated to create a mixture of gas and liquid phases. As the gas phase ascends through each tray in the column, less volatile components like nitrogen and methane undergo partial condensation. This process of repeated partial condensation occurs across multiple trays. As the gas phase rises, it becomes enriched in volatile components such as helium and hydrogen, which are collected as crude helium at the top of the column. Conversely, as the liquid phase descends, there is partial vaporization of helium, hydrogen, and other volatile components on each tray. This repeated partial vaporization leads to an increasing concentration of less volatile components like nitrogen and methane in the liquid phase, which is collected as a liquid mixture of nitrogen and methane at the bottom of the column.

For a binary mixture undergoing continuous rectification, the flow rates and compositions of the distillate and bottom product are related to the feed flow rate and composition (Fig. 4.6).

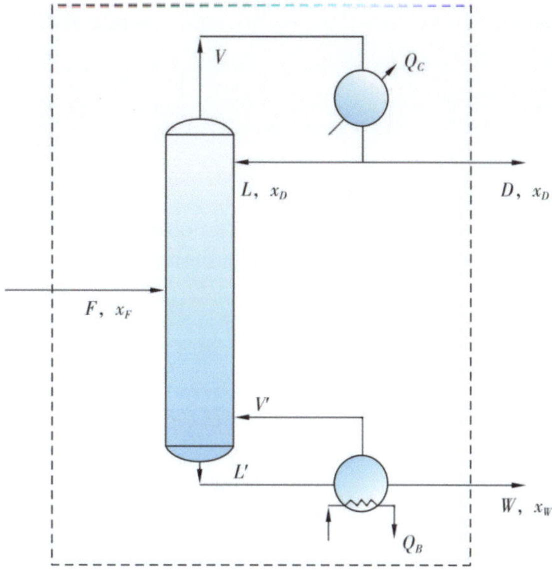

Fig. 4.6 Material balance in a rectification column

Total material balance:

$$F = D + W \tag{4.7}$$

Material balance for volatile components:

$$x_F \cdot F = x_D \cdot D + x_W \cdot W \tag{4.8}$$

where: F represents the feed flow rate of the helium-containing raw material (kmol/h); D represents the distillate flow rate at the top of the column (kmol/h); W represents the residue flow rate at the bottom of the column (kmol/h); x_F represents the mole fraction of volatile components in the feed liquid; x_D represents the mole fraction of volatile components in the distillate; x_W represents the mole fraction of volatile components in the residue.

In the rectification section, it is assumed that the mole flow rates of the vapor and liquid leaving each tray are constant, denoted as V for vapor and L for liquid; the number of theoretical trays is counted from the top down, with the vapor and liquid compositions leaving each tray represented as $y_1, x_1, y_2, x_2, \ldots, x_N$. The vapor rising from the first theoretical plate is directed to the condenser at the top of the column, where all the vapor with composition y_1 is condensed, and such condenser is referred to as a total condenser. The composition of the condensed liquid x_D is, therefore, equal to y_1. Part of this liquid D serves as the top product, while the remainder L is refluxed back into the column.

After $y_1 = x_D$, the composition x_l of the liquid descending from the first tray can be determined using the phase equilibrium relationship. To find y_2, a material balance on the top of the rectification column gives:

$$V = L + D \tag{4.9}$$

$$y_2 \cdot V = x_1 \cdot L + x_D \cdot D \tag{4.10}$$

where: L is the reflux flow rate (kmol/h).

From these equations, we can derive:

$$y_2 = x_1 \cdot L/(L + D) + x_D \cdot D/(L + D) \tag{4.11}$$

The ratio R of the reflux liquid L to the distillate D is known as the reflux ratio: $R = L/D$. Using this, Eq. (4.11) can be rewritten as (4.12):

$$y_2 = x_1 \cdot R/(R + 1) + x_D \cdot R/(R + 1) \tag{4.12}$$

With a known reflux ratio, y_2 can be calculated, and then using the phase equilibrium relationship, x_2 can be determined. This iterative process of alternating between phase equilibrium and material balance relationships allows for the calculation of the gas and liquid compositions leaving each theoretical plate down to the feed plate, at which point the operating line equation for the rectification section can be obtained.

$$y = x \cdot R/(R + 1) + x_D \cdot R/(R + 1) \tag{4.13}$$

For the stripping section, due to the presence of feed flow on the feed plate, the liquid and vapor flow rates in this section differ from those in the rectification section (represented as L and V, respectively). As shown in Fig. 4.6, we can derive the corresponding operating line equation for the stripping section.

$$y = x \cdot L'/(L' - W) + x_W \cdot W/(L' - W) \tag{4.14}$$

Rectification helium extraction offers significant economic benefits and, compared to flash evaporation, features high equipment integration and wide adaptability to raw materials, though with a more complex process flow. It falls into single-column and dual-column processes.

The process flow of single-column helium extraction is shown in Fig. 4.7 [2]. Helium-containing natural gas is pre-cooled to a specific temperature in a helium extraction heat exchanger before entering a separator for gas–liquid separation. The liquid phase, after throttling, is reheated in the helium extraction heat exchanger, pressurized with a compressor, and then transported out. The gas phase is further condensed in the helium extraction heat exchanger and fed into the middle of the helium rectification column. The liquid phase at the bottom of the helium rectification column, after reheating with a cryogenic heat exchanger and pressurization with a

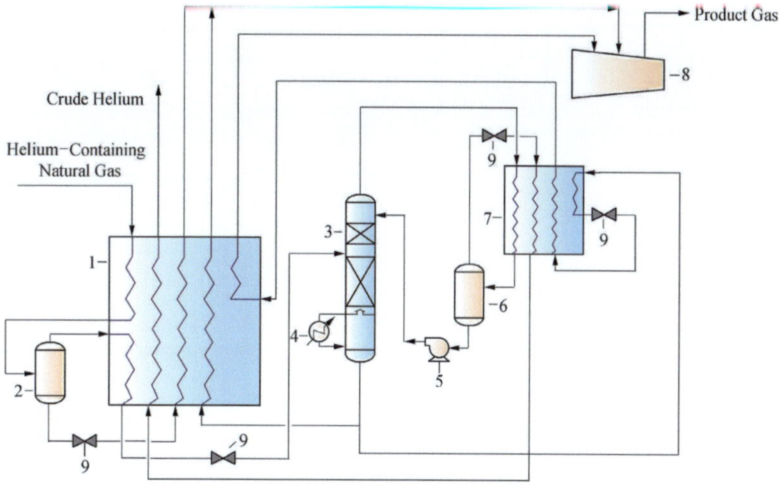

1 - Helium Extraction Heat Exchanger; 2 - Separator; 3 - Helium Rectification Column; 4 - Reboiler; 5 - Reflux Pump; 6 - Reflux Tank; 7 - Cryogenic Heat Exchanger; 8 - Natural Gas Compressor; 9 - Regulating Valve

Fig. 4.7 Flow chart of single-column helium extraction

compressor, is also transported out as product gas. The gas phase at the top of the helium rectification column is cooled in a cryogenic heat exchanger and subjected to gas and liquid separation. The liquid phase is refluxed to the top of the helium rectification column, while the gas phase, after reheating in the helium extraction heat exchanger, is exported as crude helium. The single-column process is straightforward, requires a low one-off investment, and is suitable for helium extraction from natural gas with simple compositions.

The process flow of dual-column helium extraction is shown in Fig. 4.8 [3]. Helium-containing natural gas is first cooled in a precooling heat exchanger before entering the primary rectification column. The primary crude helium at the top of the column is cooled in a cryogenic heat exchanger and then fed into the secondary rectification column. The product crude helium is separated at the top of the secondary rectification column. The majority of the liquid methane from the bottom of the primary rectification column is first passed through the precooling heat exchanger to recover part of the cooling capacity, then expanded and refrigerated before passing through the precooling heat exchanger again to recover cooling capacity. This produces medium-pressure tail gas, which is then pressurized by the expansion machine's compression side and exported as product natural gas. Another portion of the liquid methane, after throttling, serves as the cooling source for the top of the primary rectification column, then passes through the precooling heat exchanger to recover cooling capacity, and is finally pressurized by the natural gas compressor and exported as product natural gas. The cooling source at the top of the secondary rectification column is provided by nitrogen cycle refrigeration. The helium content in the crude helium can reach 60%–75%, with a helium recovery rate of over 90%.

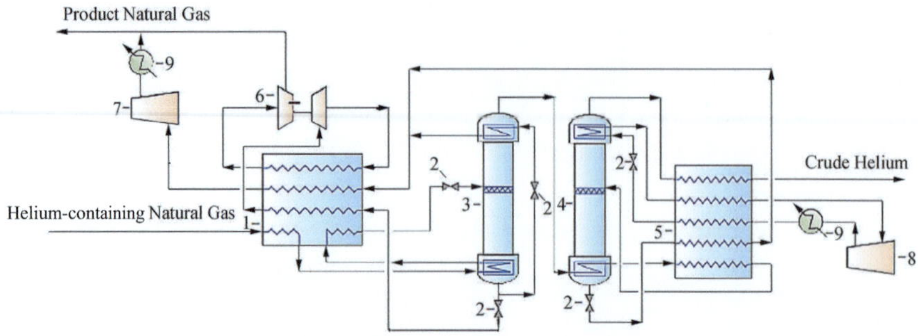

1 - Precooling Heat Exchanger; 2 - Regulating Valve; 3 - Primary Rectification Column; 4 - Secondary Rectification Column; 5 - Cryogenic Heat Exchanger; 6 - Expansion Machine; 7 - Natural Gas Compressor; 8 - Nitrogen Compressor; 9 - Cooler

Fig. 4.8 Flow chart of dual-column helium extraction

4.2 Membrane Separation Technology

Due to the higher permeability of helium in helium-containing natural gas compared to nitrogen, carbon dioxide, hydrocarbons, and other components, membrane separation technology utilizes the differences in permeability during dissolution-diffusion processes within the membrane to achieve helium concentration under the driving force of pressure differentials across the membrane. Compared to cryogenic technology, membrane separation technology does not involve phase changes and can reduce process energy consumption. Helium-containing natural gas also includes methane, carbon dioxide, hydrogen, nitrogen, and helium, among other components. The selective permeation of different components through a certain hollow fiber membrane is illustrated in Fig. 4.9.

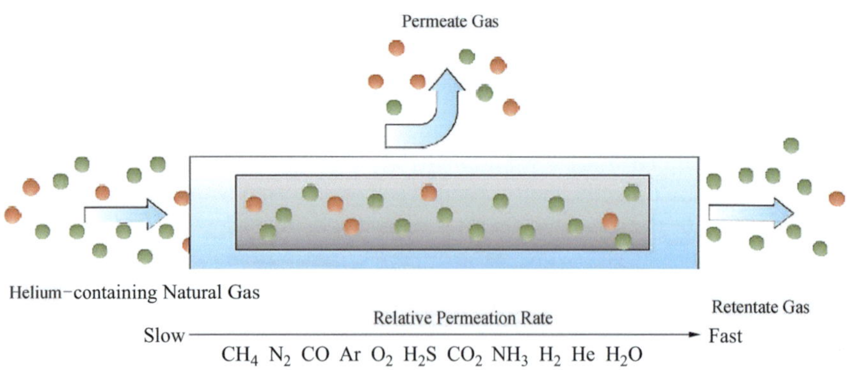

Fig. 4.9 Selectivity of different gases in a certain hollow fiber membrane

4.2.1 Principles of Membrane Separation

Membrane separation technology offers functions such as separation, concentration, purification, and refining. Its processes are simple and easy to control, making it increasingly popular in various separation applications. The applicability of membrane separation technology is depicted in Fig. 4.10. From a structural perspective, membrane materials can be classified into porous and dense membranes, each with different separation mechanisms.

Regarding porous membranes, based on the size of gas molecules and the pore size of the material, gas mixtures can be separated through mechanisms like convective flow, Knudsen diffusion, and molecular sieving, as illustrated in Fig. 4.11.

Dense membranes typically follow the solution-diffusion mechanism, which involves three main steps: solubilization, diffusion, and desorption (Fig. 4.12). The process of gas permeation through a dense membrane is described as follows:

(1) Solubilization. The gas first contacts the membrane and then dissolves on the membrane's surface.
(2) Diffusion. Due to the concentration gradient created by the dissolved gas, the gas diffuses through the membrane until it reaches the other side, during which the process is in a non-steady state.
(3) Desorption. Once the concentration gradient within the membrane aligns in a straight line along the membrane's thickness, a steady state is reached, and the gas begins to desorb from the other side of the membrane.

The permeability is commonly used to indicate the ease with which a gas can pass through a membrane, and it is a critical parameter reflecting the performance of a membrane.

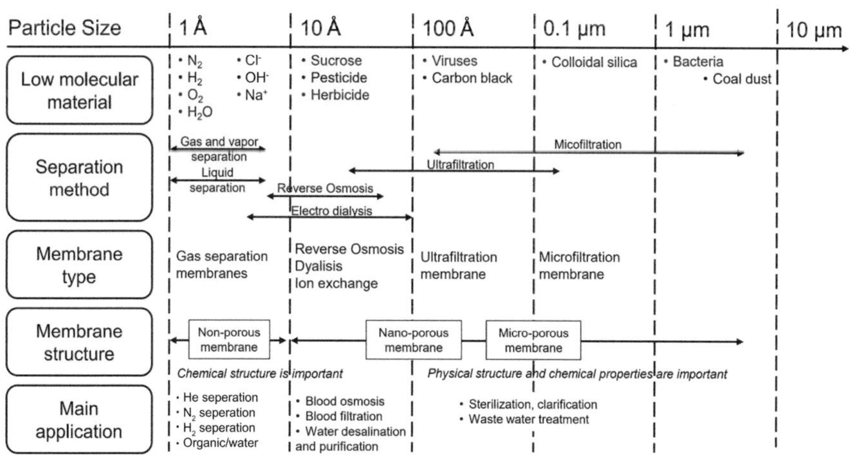

Fig. 4.10 The range of applications for membrane separation technology [4]

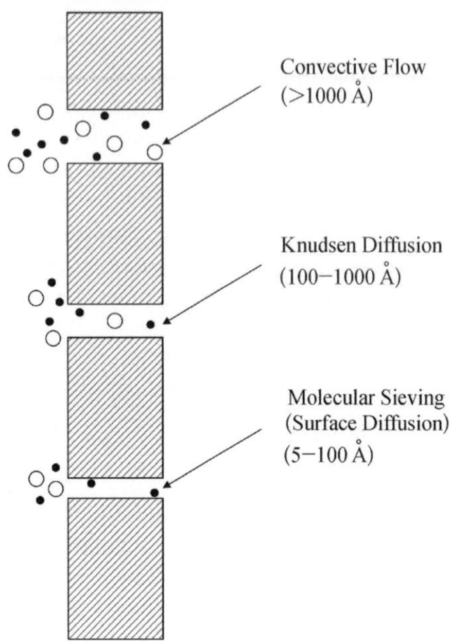

Fig. 4.11 Transport mechanism of gas molecule through porous membranes (pore flow model)

$$P = S \times D \tag{4.15}$$

In the formula: P represents the permeability; S represents the solubility coefficient (thermodynamic factor); D represents the diffusion coefficient (kinetic factor). The solubility coefficient indicates the membrane's capacity to absorb the gas, which depends on the type of dissolved gas and the polymer species of the membrane. The diffusion coefficient is defined by the volume of gas that permeates through the membrane per unit time.

The permeability is influenced by the type of gas, the chemical composition, and the molecular structure of the membrane material. When the same gas passes through different gas separation membranes, the permeability mainly depends on the diffusion coefficient of the gas in the membrane. Conversely, when different gases pass through the same membrane, the permeability primarily hinges on the solubility coefficient of each gas in the membrane.

Gas permeability refers to the flow rate of gas through a unit area of the membrane. The size of the gas permeation flux directly affects the efficiency and cost of membrane permeation. The expression for gas permeability is as follows:

$$P = \frac{Q}{t \cdot \Delta p \cdot A} \tag{4.16}$$

where: P is the gas permeability, typically expressed in Gas Permeation Unit (GPU), with 1 GPU = 10^{-6} cm^3(STP)/(cm^2·s·cmHg) = 3.348×10^{-10} mol/(m^2·s·Pa); Q is

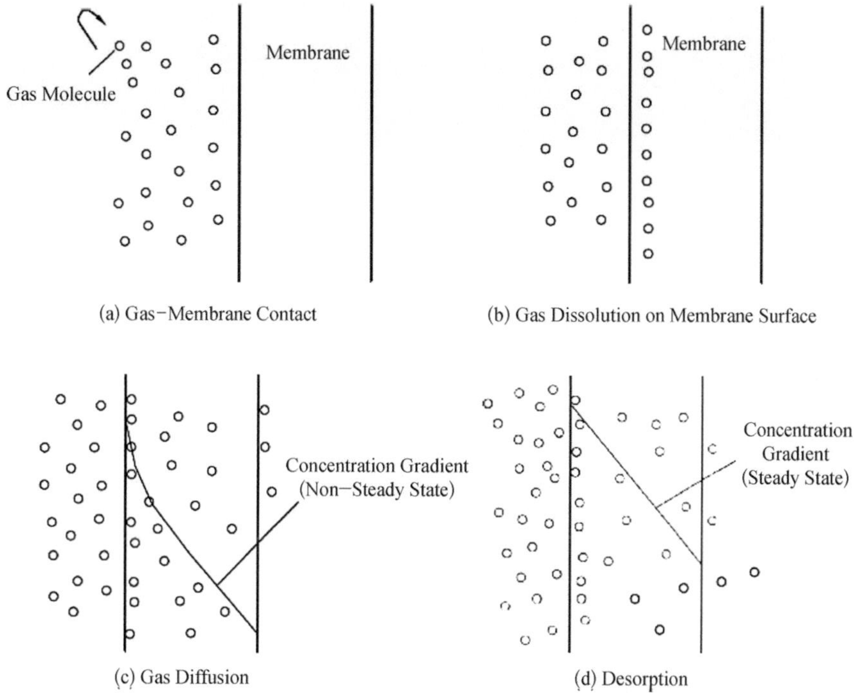

Fig. 4.12 Transport mechanism of gas molecule through dense membranes (solution-diffusion model)

the gas permeation flux, in mol; t is the time of permeation, in second (s); Δp is the pressure differential, in Pascal (Pa); A is the membrane area, in square meter (m^2).

Generally, the selectivity of a membrane is used to measure its separation performance, while the separation coefficient evaluates the membrane's selective separation performance for gases. The expressions for the gas separation coefficient are as follows:

$$\alpha_{A/B} = \frac{\left(\frac{C_{A'}}{C_{B'}}\right)_{\text{Permeate Gas}}}{\left(\frac{C_A}{C_B}\right)_{\text{Feed Gas}}} = \frac{p_{A'}/p_{B'}}{p_A/p_B} \tag{4.17}$$

where: $\alpha_{A/B}$ is the separation coefficient; c represents the concentration of the gas component, in parts per million by volume (ppmV); p represents the pressure of the gas component, in Pa.

The membrane used to separate helium is a dense membrane. Membrane separation takes advantage of the differences in the permeation rates of different gas molecules within the membrane. Driven by pressure, molecules with relatively faster permeation rates accumulate on the permeate side, while molecules with relatively slower permeation rates accumulate on the retentate side, thereby achieving the

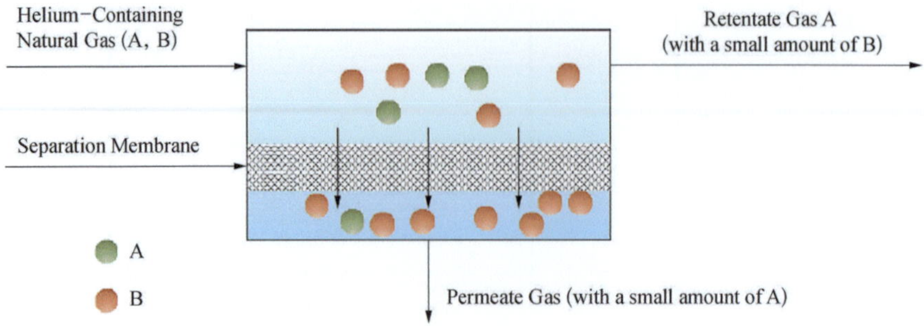

Fig. 4.13 Schematic diagram of the membrane separation process

separation of gas mixtures. Generally, membranes are permeable to gases, although different gases have varying permeation rates. Under normal conditions, molecules with relatively fast permeation rates pass through the membrane more readily, becoming the crude permeate gas, while molecules with slower permeation rates remain more in the feed gas side and become the retentate gas, as illustrated in Fig. 4.13. The relative speed of molecular permeation is specific to different gas compositions. For instance, compared to methane, helium has a faster permeation rate, whereas methane has a slower rate. In contrast, compared to helium, water vapor has a faster permeation rate, while helium has a slower rate. Thus, the relative permeation rates are primarily determined by the specific gas system under consideration.

4.2.2 Helium Extraction by Membrane Separation

Helium extraction by Membrane separation typically does not involve phase changes and can effectively increase helium concentration. The separation performance of membrane systems largely depends on process configuration. Currently, polyimide hollow fiber membranes are commonly used to concentrate helium from helium-containing natural gas. Prior to membrane entry, solid and liquid impurities in the raw natural gas should be removed, with a recommended filtration precision of more than 1 μm. Furthermore, different membrane products have varying tolerances for different components. Typically, the feed natural gas entering the membrane should have NH_3 content less than 1×10^{-6} (by volume), methanol content less than 200×10^{-6} (by volume), and H_2S partial pressure less than 7 bar. The membrane operating temperature should be higher than the hydrocarbon dew point by 25 °C, and the hydrocarbon dew point of the feed gas should be less than 40 °C (under operating pressure). The performance of helium separation by membranes is sensitive to temperature, so it is advisable to install a heater before entering the membrane to control the temperature of the raw natural gas and ensure efficient helium separation.

The pressure difference of helium between the feed side and the permeate side serves as the driving force for the helium extraction process via membrane separation. Considering the constraints of helium recovery conditions and the differences in the pressure and concentration of raw natural gas, specific helium pressure differences can be generated by using compressors on the feed side or vacuum pumps on the permeate side to obtain sufficient driving force. This approach aims to improve separation efficiency, reduce energy consumption, and lower gas processing costs. Figure 4.14 illustrates different process flows to generate transmembrane driving forces in single-stage membrane separation.

According to the number of membrane stages, membrane separation technology can be classified into single-stage membrane helium extraction processes and multi-stage ones. In single-stage membrane separation, raw natural gas passes through a membrane separator to obtain retentate gas and permeate gas, where helium is concentrated in the permeate gas. Typically, single-stage membrane separation can enrich helium from helium-containing natural gas by 8–12 times, as shown in Fig. 4.14.

To increase the helium concentration or recovery rate from natural gas, a multi-stage membrane separation system is required. Therefore, in addition to single-stage membrane separation, there are permeate-side multi-stage membrane separation processes and retentate-side multi-stage ones for the purpose of extracting helium. Increasing the number of membrane separation stages on the permeate side can

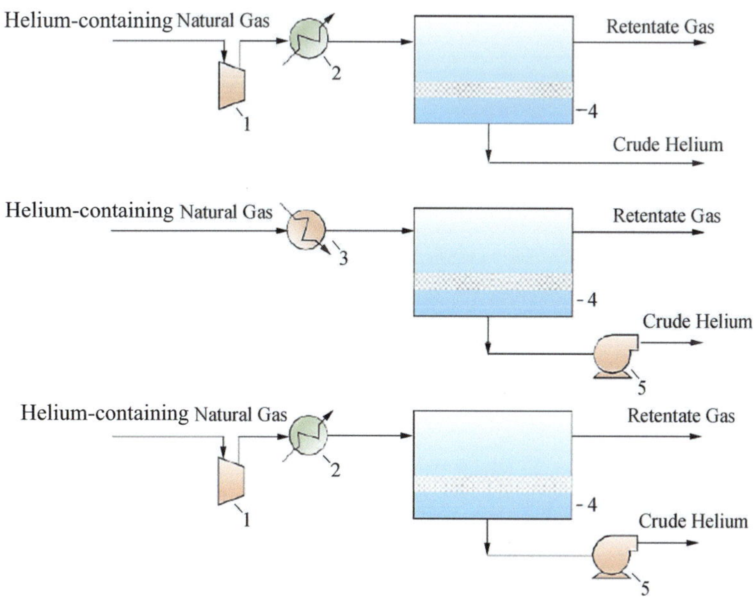

Fig. 4.14 Different process flows for generating driving forces in single-stage membrane separation

further boost the helium concentration in the permeate gas. As depicted in Fig. 4.15, multiple stages of helium enrichment in the permeate gas increase the loss of helium in the retentate gas, thereby reducing the helium recovery rate. Increasing the number of membrane separation stages on the retentate-side to recover helium from the retentate gas can further ramp up the helium recovery rate, albeit with a slight reduction in helium concentration in the permeate gas, as shown in Fig. 4.16. Balancing the helium recovery rate and concentration in membrane separation requires comprehensive technical and economic evaluation based on factors such as helium content in the raw natural gas, helium recovery rate, and processing scale. It can also be combined with cryogenic technology and pressure swing adsorption technology for crude helium extraction.

The process flow of two-stage, three-section membrane separation improved by a certain factory is shown in Fig. 4.17. Helium-containing natural gas is first heated to 50 °C in a heater before entering the first-stage membrane separator for permeation and separation. The permeate gas from the first-stage membrane separator is mixed with that from the second-stage second-section membrane separator, and then pressurized in the intermediate compressor before being sent to the second-stage first-section membrane separator. Compared to the raw natural gas, this process records approximately a 500-fold increase in helium concentration in crude helium and an 85% recovery rate.

Membrane separation technology is suitable for small-scale helium extraction from natural gas, offering advantages such as high separation efficiency, simple operation, and relatively low energy consumption. However, limited by the helium

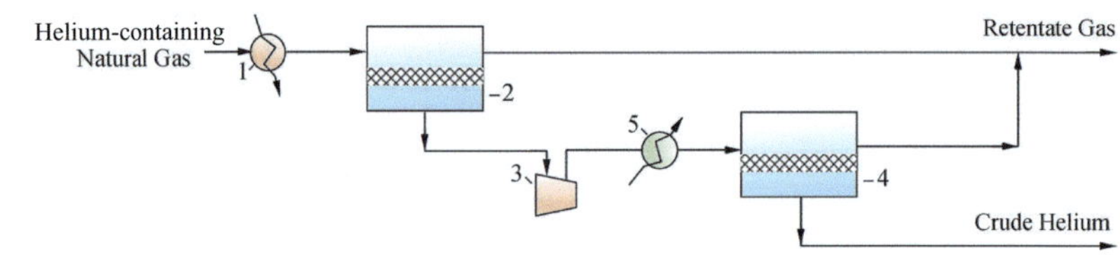

1 - Heater; 2 - First-Stage Membrane Separator; 3 - Intermediate Compressor; 4 - Second-Stage Membrane Separator; 5 - Cooler

Fig. 4.15 Schematic diagram of multi-stage membrane separation for permeate gas

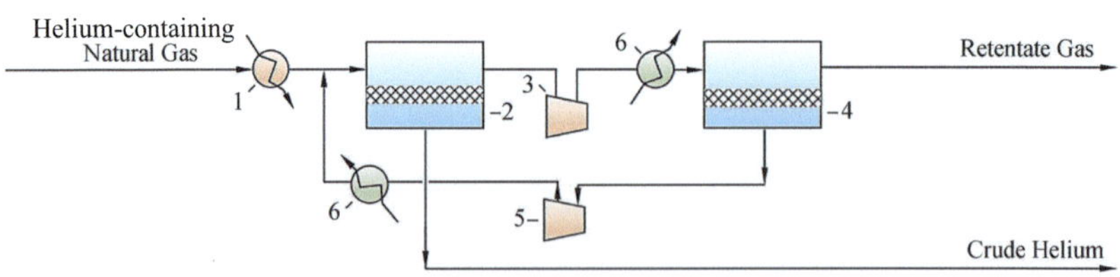

1 - Heater; 2 - First-Stage Membrane Separator; 3 - Intermediate Compressor; 4 - Second-Stage Membrane Separator; 5 - Recovery compressor; 6 - Cooler

Fig. 4.16 Schematic diagram of multi-stage membrane separation for retentate gas

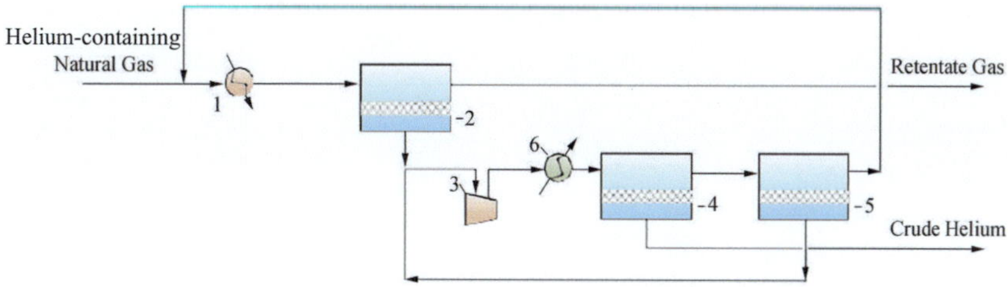

1 - Heater; 2 - First-Stage Membrane Separator; 3 - Intermediate Compressor; 4 - Second-Stage First-Section Membrane Separator; 5 - Second-Stage Second-Section Membrane Separator; 6 - Cooler

Fig. 4.17 Process flow of two-stage, three-section membrane separation technology

partial pressure in helium-containing natural gas, its helium recovery rates range from 60% to 90%, requiring further improvements in reliability and stability. Membrane components, whose lifespan is influenced by impurities such as composition and dust in helium-containing natural gas, need to be replaced at fixed periods.

4.3 Combined Membrane Separation and Cryogenic Technology

Currently, membrane separation technology exhibits high separation efficiency, easy operation, and relatively low energy consumption. However, it has drawbacks such as low helium recovery rates and the need for further improvements in reliability and stability, making it suitable for small-scale operations and low-helium-content natural gas. On the other hand, cryogenic technology is known for its high helium recovery rates, reliability, and stability, making it more suitable for large-scale operations and high-helium-content natural gas, despite its relatively higher energy consumption. Both helium extraction technologies have their advantages and disadvantages. By combining membrane separation and cryogenic technologies, the most economical helium extraction process can be achieved, enabling the production of high-purity helium from low-helium natural gas while minimizing energy consumption and operating costs.

The process flow of the first-stage membrane separation and dual-column cryogenic helium extraction is shown in Fig. 4.18. The helium-containing natural gas is directed into the membrane separation unit, where the permeate gas is transported externally as product gas, and the concentrated helium-rich permeate gas is compressed and subjected to decarbonization and dehydration processes. The purified helium-rich natural gas is then precooled before being introduced into the middle section of the primary rectification column. The overhead gas from the primary rectification column undergoes condensation and gas–liquid separation. The resulting liquid phase is returned to the top of the primary rectification column as reflux, while

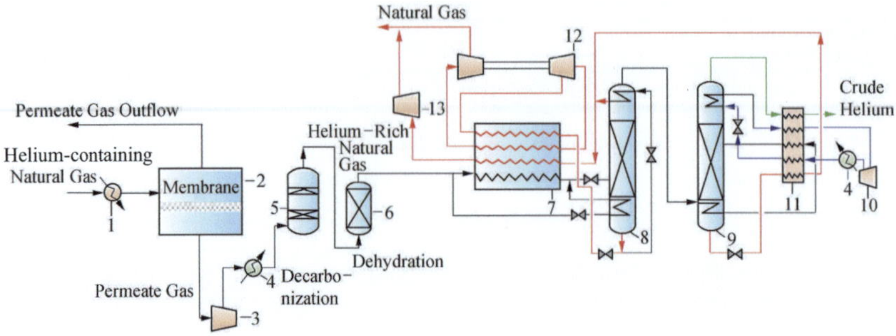

1 - Heater; 2 - Membrane Separator; 3 - Compressor; 4 - Cooler; 5 - Decarbonization Tower; 6 - Dehydration Tower; 7 - Precooling Heat Exchanger; 8 - Primary Rectification Column; 9 - Secondary Rectification Column; 10 - Nitrogen Compressor; 11 - Cryogenic Heat Exchanger; 12 - Expander; 13 - Natural Gas Compressor

Fig. 4.18 Process flow of first-stage membrane separation and dual-column cryogenic helium extraction

the gas phase is further cooled and sent to the secondary rectification column for rectification. The liquid hydrocarbons at the bottom of the primary rectification column are divided into two streams. One stream is sequentially reheated through a precooling heat exchanger, pressurized by an expander, and transported as high-pressure product gas. The other stream is throttled to the top condenser of the primary rectification column, providing cooling before being reheated and transported as low-pressure product gas. The overhead gas from the secondary rectification column undergoes condensation and gas–liquid separation. The resulting liquid phase is returned to the top of the secondary rectification column as reflux, while the gas phase is reheated and transported as crude helium. The liquid hydrocarbons at the bottom of the secondary rectification column are reheated and pressurized before being transported as low-pressure product gas. A nitrogen refrigeration cycle is employed to provide cooling for the system. The primary rectification column achieves the separation of helium from methane and increases helium recovery rates. The secondary rectification column separates helium from a large amount of nitrogen and the remaining methane and other hydrocarbons, resulting in crude helium with a helium content of up to 65%.

4.4 Multi-product Co-production Technology

When the helium content in natural gas is low, the cost of using cryogenic technology solely for helium extraction from natural gas runs high. Therefore, co-producing multiple products while extracting helium using cryogenic methods is considered in an attempt to enhance economic benefits. This includes co-production scenarios such as crude helium extraction with LNG, ethane, and a combination of both, among others. In industrial applications, the process for extracting crude helium from natural gas must be selected based on factors such as gas source characteristics, engineering requirements, and economic feasibility.

Helium-containing natural gas can utilize a cryogenic helium extraction process co-producing LNG, producing LNG products alongside helium. For raw natural gas containing helium along with certain amounts of ethane, propane, butane, and pentane, processes involving ethane recovery and cryogenic helium extraction can be used to produce helium and ethane products, or alternatively, co-produce helium, ethane, and LNG products.

4.4.1 Crude Helium Extraction Co-production with LNG

The cryogenic helium extraction co-production with LNG utilizes a process of dual-column cryogenic rectification followed by expansion [5]. The helium-containing natural gas is first processed through a decarbonization tower to remove the carbon dioxide, and then passes through a dehydration tower to remove water from the gas. After precooling in a plate-fin heat exchanger, the gas sequentially enters a primary rectification column to remove methane and other hydrocarbons, followed by a secondary rectification column to remove nitrogen. The crude helium from the top of the secondary rectification column is recovered for cooling capacity in a plate-fin heat exchanger before proceeding downstream. A portion of the liquid hydrocarbons from the bottom of the primary rectification column undergoes cooling capacity recovery in a plate-fin heat exchanger, and then after expansion and further cooling capacity recovery, it is pressurized for export as product gas. Another portion of the liquid hydrocarbons is throttled and then enters the condenser at the top of the primary rectification column. The gas phase undergoes cooling capacity recovery in a plate-fin heat exchanger for export as product gas, while the liquid phase is collected as LNG product. External refrigerant circulation can be used to provide cooling capacity to the plate-fin heat exchanger to boost LNG production (Fig. 4.19).

4.4.2 Crude Helium Extraction Co-production with Ethane

When the helium-containing natural gas involves a high proportion of ethane, crude helium extraction co-production with ethane can be employed to recover ethane in the crude helium extraction process and thus to enhance economic viability. The process flow is illustrated in Fig. 4.20.

The helium-containing natural gas undergoes precooling in an ethane recovery heat exchanger before entering a separator for gas–liquid separation. The liquid phase, after throttling, flows to the bottom of the demethanizer, while the gas phase splits into two streams. One stream undergoes further condensation to liquid phase and then ascends to the top of the demethanizer, and the other stream passes through an expander before entering the middle of the demethanizer. The hydrocarbon liquid

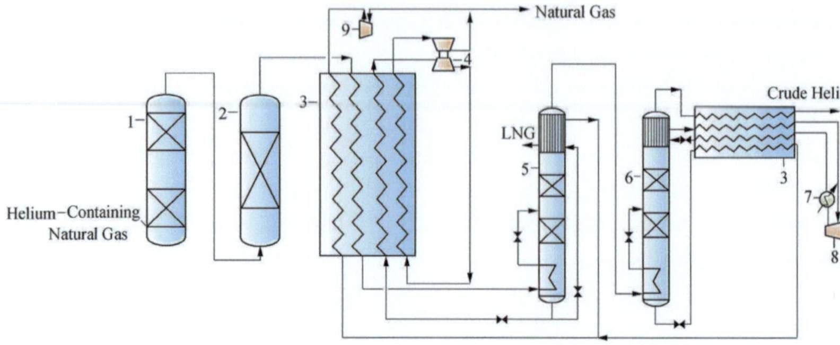

1 - Decarbonization Tower; 2 - Dehydration Tower; 3 - Plate-Fin Heat Exchanger; 4 - Expander; 5 - Primary Rectification Column; 6 - Secondary Rectification Column; 7 - Cooler; 8 - Nitrogen Compressor; 9 - Natural Gas Compressor

Fig. 4.19 Process flow of crude helium extraction co-production with LNG

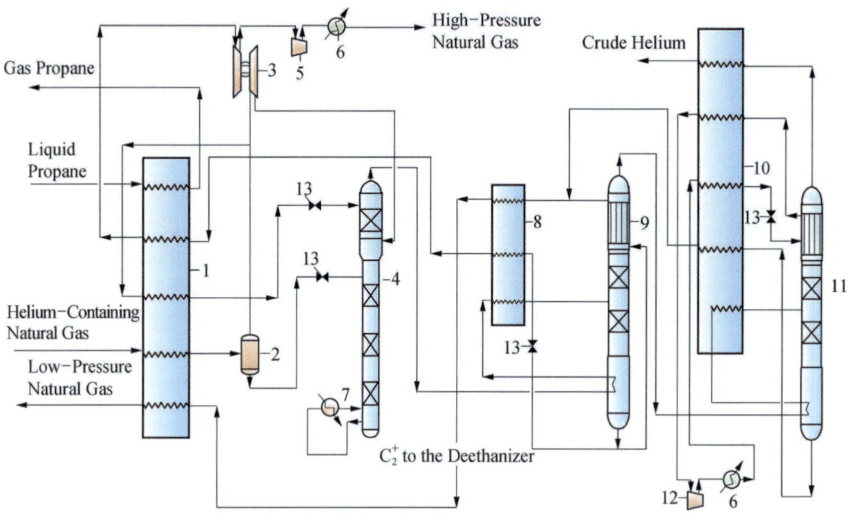

1 - Ethane Recovery Heat Exchanger; 2 - Separator; 3 - Expander; 4 - Demethanizer; 5 - Natural Gas Compressor; 6 - Cooler; 7 - Reboiler; 8 - Precooling Heat Exchanger; 9 - Primary Rectification Column; 10 - Cryogenic Heat Exchanger; 11 - Secondary Rectification Column; 12 - Nitrogen Compressor; 13 - Regulating Valve

Fig. 4.20 Process flow of crude helium extraction co-production with ethane

from the bottom of the demethanizer flows to the deethanizer for separation of ethane and other products. The gas from the top of the demethanizer undergoes precooling before entering the primary concentration column for rectification. The gas from the top of the primary concentration column undergoes condensation to achieve gas–liquid separation, with the liquid phase returning to the top of the primary concentration column as reflux, and the gas phase passing through condensation before entering the secondary concentration column for rectification. The liquid hydrocarbons from the bottom of the primary concentration column are divided into two streams. One stream is sequentially reheated and then transported as high-pressure product gas, while the other stream, after throttling, enters the condenser at the top of the primary concentration column for cooling before being reheated to ambient temperature and transported as low-pressure product gas. The liquid hydrocarbons at the bottom of the secondary concentration column are reheated and pressurized before being transported as low-pressure product gas. The overhead gas from the secondary concentration column undergoes condensation and gas–liquid separation. The resulting liquid phase is returned to the top of the secondary concentration column as reflux, while the gas phase is reheated and transported as crude helium. A nitrogen refrigeration cycle is employed to provide cooling for the system.

4.4.3 Crude Helium Extraction Co-production with Ethane and LNG

When helium-containing natural gas has a high content of ethane, to enhance the economic feasibility of crude helium extraction, co-production of ethane and LNG can be implemented simultaneously during the crude helium extraction process, as illustrated in Fig. 4.21.

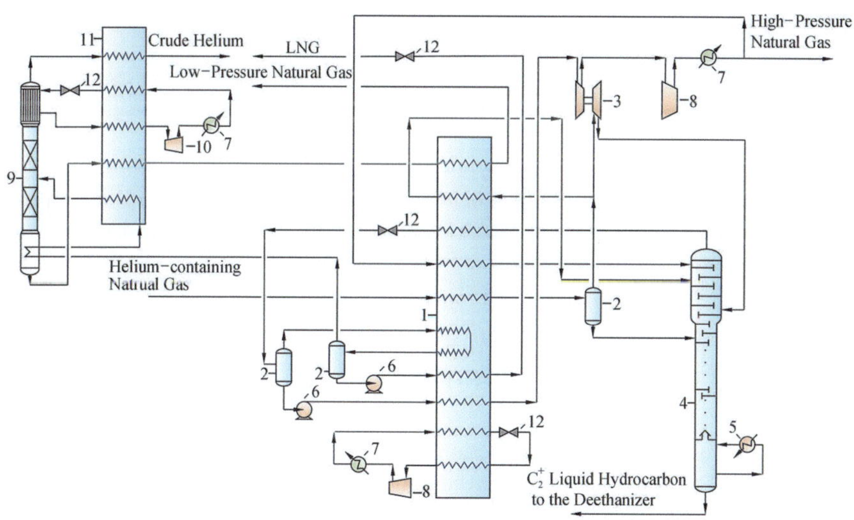

1 - Ethane Recovery and Liquefaction Heat Exchanger; 2 - Separator; 3 - Expander; 4 - Demethanizer; 5 - Reboiler; 6 - Liquid Hydrocarbon Pump; 7 - Cooler; 8 - Mixed Refrigerant Compressor; 9 - Helium Concentration Column; 10 - Nitrogen Compressor; 11 - Helium Extraction Heat Exchanger; 12 - Regulating Valve

Fig. 4.21 Process flow of crude helium extraction co-production with ethane and LNG

The helium-containing natural gas enters the ethane recovery and liquefaction heat exchanger for precooling before entering the cryogenic separator for gas–liquid separation. The liquid phase, after throttling, enters the bottom of the demethanizer, while the gas phase is split into two streams. One stream undergoes further condensation and ascends to the top of the demethanizer, and the other stream passes through an expander before entering the middle of the demethanizer. The liquid hydrocarbons from the bottom of the demethanizer flow to the deethanizer for separation of ethane. The gas from the top of the demethanizer undergoes condensation and then enters the first-stage separator for gas–liquid separation. The liquid phase is reheated and pressurized before being transported as product gas, while the gas phase undergoes further condensation and throttling before entering the second-stage separator for gas–liquid separation. The liquid phase, after further cooling, is processed as LNG product, while the gas phase, after precooling, enters the helium concentration column for rectification. The liquid hydrocarbons from the bottom of the helium rectification column are recovered for cooling capacity and pressurized before being transported as product gas. The overhead gas from the helium rectification column undergoes condensation and gas–liquid separation. The resulting liquid phase is returned to the top of the helium rectification column as reflux, while the gas phase is reheated and transported as crude helium. Mixed refrigerant and nitrogen refrigeration cycles are used to provide cooling for the system. The nitrogen refrigeration cycle provides cooling for the helium extraction heat exchanger and the condenser at the top of the helium concentration column. The ethane recovery rate of this process can reach above 93%, and the helium recovery rate can clock up over 90%.

4.5 Multiple Gas Source Technology

In processes such as LNG and synthetic ammonia production using helium-containing natural gas as the feedstock, helium is enriched as an inert gas. Therefore, boil-off gas (BOG) and exhaust gases from synthetic ammonia can also serve as sources for crude helium extraction.

4.5.1 BOG Crude Helium Extraction

BOG is generated during the production, storage, and transportation of LNG as a result of heat exchange with the surrounding environment, causing a temperature increase in the LNG and resulting in partial vaporization of the liquid. BOG mainly consists of methane, nitrogen, helium, hydrogen, etc. The helium content in BOG ranges from 0.5% to 3%, which is about 20 times higher in helium rectification compared to the feed gas for LNG production. The use of BOG for helium extraction has significant potential. Due to the relatively simplified composition of BOG, multi-stage membrane separation or single-column cryogenic processes can be employed for helium extraction.

1. BOG Multi-stage Membrane Separation Process

The typical process flow of BOG multi-stage membrane separation for helium extraction is shown in Fig. 4.22. The retentate gas of BOG is exported after passing through the first-stage membrane separator, the helium-rich permeate gas undergoes pressurization before entering the second-stage first-section membrane separator, and the permeate gas of the second-stage first-section membrane separator is transported as crude helium. To boost helium recovery, the retentate gas from the second-stage first-section membrane separator enters a second-stage second-section membrane separator for further separation. The permeate gas from the second-stage second-section membrane separator is mixed with that from the first-stage membrane separator, pressurized, and then recycled back to the inlet of the second-stage first-section membrane separator. The retentate gas from the second-stage second-section membrane separator is mixed with BOG and recycled back to the inlet of the first-stage membrane separator.

2. BOG Single-Column Cryogenic Helium Extraction Process

The process flow of BOG single-column cryogenic helium extraction is depicted in Fig. 4.23. The BOG (with a helium content of 2%–3%) is first compressed after reheating and then proceeds for condensation before entering the helium rectification column. The liquid hydrocarbons from the bottom of the helium concentration column are reheated for further processing in the next step, while the gas at the top is further cooled for gas–liquid separation. The liquid phase serves as reflux for the helium concentration column, while the gas phase, after reheating, is processed as crude helium (with helium content reaching up to 89%) for the next step. A nitrogen refrigeration cycle is employed to provide cooling for the system.

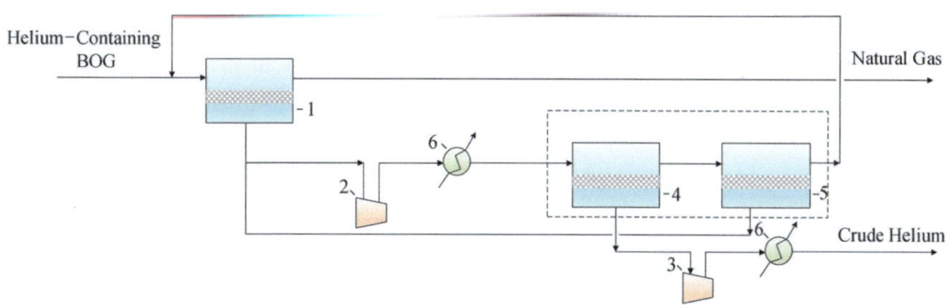

1 - First-Stage Membrane Separator; 2 - Intermediate Compressor; 3 - Permeate Gas Compressor; 4 - Second-Stage First-Section Membrane Separator; 5 - Second-Stage Second-Section Membrane Separator; 6 - Cooler

Fig. 4.22 Flow chart of BOG multi-stage membrane separation for helium

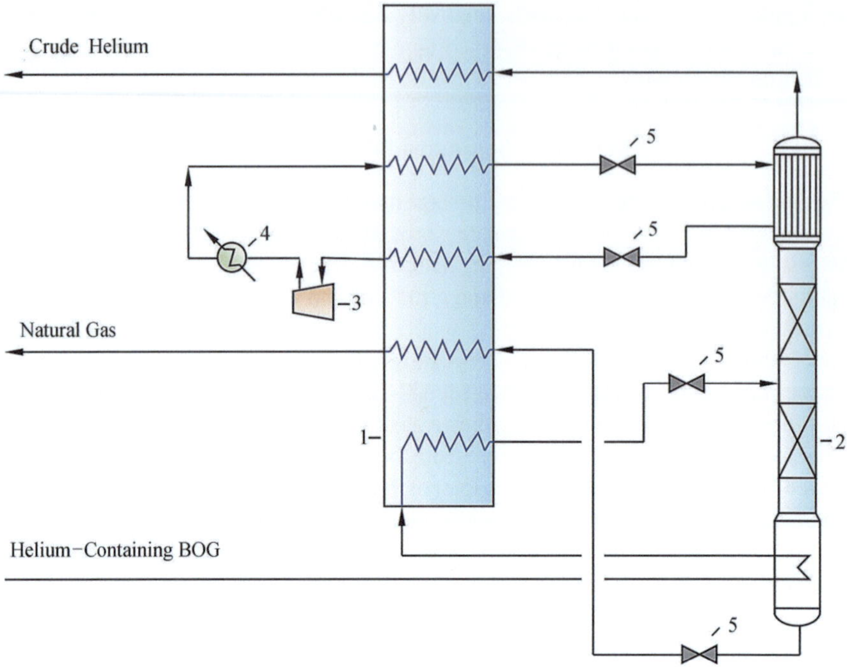

1 - Heat Exchanger; 2 - Helium Concentration Column; 3 - Nitrogen Compressor; 4 - Cooler; 5 - Regulating Valve

Fig. 4.23 Flow chart of BOG single-column cryogenic helium extraction

4.5.2 *Exhaust Gas Crude Helium Extraction*

In chemical processes, all or part of the low-grade gas components that cannot be further used are continuously or periodically discharged from production units. Such gas is known as exhaust gas. Exhaust gas originates from side reactions in chemical processes or residual components from the use of mixed gas raw materials. For example, large amounts of exhaust gases are discharged during processes such as synthetic ammonia and methanol production. In chemical plants using helium-containing natural gas as feedstock, such as in synthetic ammonia production, helium, which does not participate in chemical reactions, gradually accumulates in the synthetic recycle gas. The helium content in exhaust gas is higher than that in the raw natural gas, making it valuable for extraction. In recent years, with the scaling up of chemical plants, comprehensive utilization of exhaust gases has gained increasing attention.

Synthetic ammonia purge gas contains hydrogen, helium, nitrogen, argon, and trace amounts of ammonia, carbon dioxide, water, methane, and neon. Typically, the helium content is not less than 0.2%, while the hydrogen content can exceed 50%. Therefore, the key to helium recovery from synthetic ammonia exhaust gas lies in

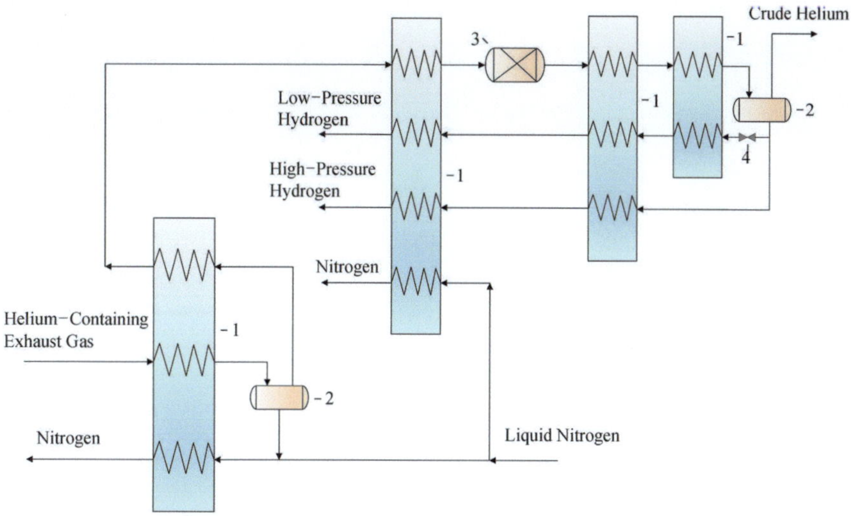

1 - Heat Exchanger; 2 - Separator; 3 - Adsorption Purifier; 4 - Regulating Valve

Fig. 4.24 Process flow of crude helium extraction using low-temperature adsorption and hydrogen condensation

the removal of a large amount of hydrogen from the mixed gas. Crude helium can be extracted using low-temperature adsorption and hydrogen condensation processes, with a typical process flow shown in Fig. 4.24.

The purified helium-containing purge gas first undergoes condensation of nitrogen components in a heat exchanger, followed by gas–liquid separation. The nitrogen-rich liquid phase, after reheating, is vented to the atmosphere, while the gas phase undergoes further condensation of most of the hydrogen, followed by gas–liquid separation. The nitrogen-rich liquid phase returns for reheating before moving on to the next stage for processing, while the gas phase is processed in the next step as crude helium subjected to concentration. Liquid nitrogen is used to provide cooling for the system.

References

1. Rufford TE, Chan KI, Huang SH, et al. A review of conventional and emerging process technologies for the recovery of helium from natural gas. Adsorpt Sci Technol. 2014;32(1):49–72.
2. Mokhatab S, Poe WA, Mak JY. Handbook of natural gas transmission and processing: principles and practices. Gulf Professional Publishing; 2018.
3. Long ZB, Ju YL, Zhong ZL, et al. Investigation and research on technology of helium extraction from natural gas. Nat Gas Oil. 2009;27(4):28–31 (in Chinese).
4. Nath A, Chakraborty S, Bhattacharjee C. Bioreactor and enzymatic reactions in bioremediation. In: Microbial biodegradation and bioremediation. New York: Elsevier; 2014. pp. 455–495.
5. Li JF, He LL, Chai LH. Present situation and suggestion of helium extraction from natural gas. Chem Eng Oil Gas. 2018;47(4):41–44 (in Chinese).

Chapter 5
Crude Helium Refining Technologies

Xuan Zhou and Yingyong Zou

The crude helium extracted from helium-containing natural gas has a helium content lower than 97.5%, with small amounts of hydrocarbons, nitrogen, hydrogen, and trace amounts of neon, among other impurities. To make higher-purity helium products, these impurities must be removed to achieve the required purity of the helium product. The process of removing impurities from crude helium to obtain helium products is known as helium refining technology, which generally includes catalytic dehydrogenation or adsorptive dehydrogenation technology for hydrogen removal, pressure swing adsorption or temperature swing adsorption technology for removing nitrogen and hydrocarbon impurities, as well as low-temperature adsorption or low-temperature pressure swing adsorption technology for removing neon and other impurities (Fig. 5.1). Different helium resources and extraction methods can result in varying compositions and concentrations of impurities in crude helium. Therefore, the process flow and technology combination for crude helium refining need to be determined based on the specific composition and concentration of impurities in crude helium, as well as the quality requirements of the helium product.

5.1 Catalytic Dehydrogenation Technology

In helium extraction from natural gas, the ratio of helium to hydrogen in the feed gas typically ranges from 2:1 to 15:1, resulting in crude helium with relatively low hydrogen content. Due to the close molecular size and condensation temperature of

X. Zhou (✉)
China Petroleum Engineering & Construction Corporation Southwest Company, Chengdu, China
e-mail: zhouxuan_sw@cnpc.com.cn

Y. Zou
PetroChina Tarim Oilfield Company, Korla, Xinjiang, China
e-mail: zouyy-tlm@petrochina.com.cn

L. Pu (ed.), *Helium Engineering Technologies*,
https://doi.org/10.1007/978-981-96-4086-7_5

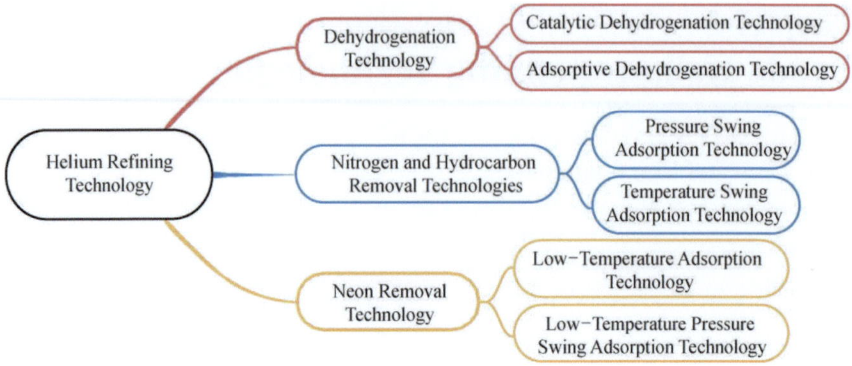

Fig. 5.1 Technology composition of crude helium refining

hydrogen and helium, conventional methods such as adsorption, membrane separation, and cryogenic rectification struggle to completely separate hydrogen from helium. In industrial practice, catalytic dehydrogenation technology is often used to remove hydrogen by injecting a slight excess of oxygen into crude helium, utilizing a catalytic hydrogen–oxygen reaction to achieve the required hydrogen content in the product standards.

5.1.1 *Principle*

Catalytic dehydrogenation technology is based on the chemical stability of helium. By injecting oxygen (or air) into crude helium, hydrogen reacts with oxygen through catalytic oxidation to generate water for the purpose of hydrogen removal. The process ensures that the hydrogen content in the dehydrogenated crude helium does not exceed 1 mL/m^3 (ppmV). The main equations are as follows:

$$2H_2 + O_2 = 2H_2O, \quad \Delta H = -571.6\,\text{kJ/mol} \tag{5.1}$$

The most common catalysts for this process are metal oxides, including iron oxide (Fe_2O_3), chromium oxide (Cr_2O_3), zinc oxide (ZnO), and magnesium oxide (MgO), as well as metals like silver (Ag), nickel (Ni), platinum (Pt), and palladium (Pd), with Pd being the most widely used.

To secure the reliability of the catalytic dehydrogenation unit, multiple stages of catalytic oxidation reactors are required in general, and each reactor's temperature is controlled to avoid excessive heat, thereby maintaining catalyst activity and ensuring a controlled reaction. Additionally, a catalytic desulfurization protection bed is usually placed before the catalytic oxidation reactors to prevent sulfur poisoning, which can deactivate the catalyst. Since dehydrogenation is an exothermic reaction, water coolers are installed after each stage of the catalytic oxidation reactor to keep down the temperature.

Principle of desulfurization: At certain temperatures, organic sulfur is catalytically converted into hydrogen sulfide (H_2S) by the desulfurization catalyst. The H_2S is then absorbed by active components. Desulfurization catalysts may use new iron-manganese-zinc composite metal oxides with dual functionality, which can remove both H_2S and organic sulfur.

Organic sulfur conversion equation:

$$COS + H_2 = H_2S + CO \tag{5.2}$$

$$CS_2 + 4H_2 = 2H_2S + CH_4 \tag{5.3}$$

$$RSH + H_2 = H_2S + RH \tag{5.4}$$

$$RSH\ (\text{mercaptan}) + H_2 = H_2S + RH \tag{5.5}$$

Hydrogen sulfide absorption equation:

$$H_2S + ZnO = ZnS + H_2O \tag{5.6}$$

$$H_2S + MnO = MnS + H_2O \tag{5.7}$$

$$2H_2S + Fe_2O_3 + H_2 = 2FeS + 3H_2O \tag{5.8}$$

Firstly, hydrogen in crude helium catalyses an oxidation reaction to produce water. Then after cooling, the gas-phase water is converted to free water, which is then removed by a separator. Saturated water in crude helium is typically removed using molecular sieve adsorption, similar to dehydration processes in natural gas treatment. The molecular sieves can be regenerated by heating, allowing for cyclic adsorption.

To ensure catalytic oxidative dehydrogenation, an excess amount of oxygen is injected into crude helium. The injected oxygen usually comes from high-purity sources (99.995% pure oxygen or industrial oxygen) to minimize the introduction of impurities into crude helium. For systems requiring larger amounts of oxygen, air may also be injected.

5.1.2 *Typical Process Flow*

Take a typical three-stage catalytic dehydrogenation process as an example, as shown in Fig. 5.2.

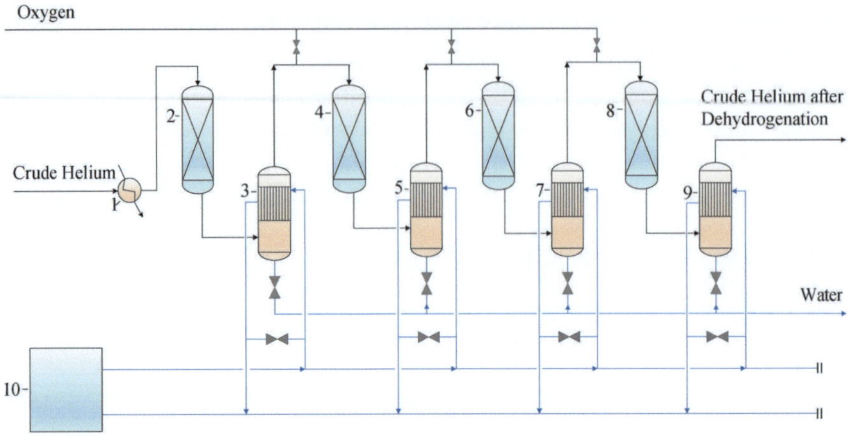

1 - Desulfurization Heater; 2 - Desulfurization Tank; 3 - Water Cooler; 4 - First-stage Dehydrogenation Reactor; 5 - First-stage Water Cooler and Separator; 6 - Second-stage Dehydrogenation Reactor; 7 - Second-stage Water Cooler and Separator; 8 - Third-stage Dehydrogenation Reactor; 9 - Third-stage Water Cooler and Separator; 10 - Water Cooler

Fig. 5.2 Process flow of three-stage catalytic dehydrogenation

Firstly, crude helium undergoes heating and then enters the desulfurization tank, where the total sulfur content is reduced to no more than 0.1 mL/m^3 (ppmV) to prevent poisoning of the subsequent dehydrogenation catalysts. Secondly, the crude helium after desulfurization is cooled to 80–120 °C using a water cooler, injected with oxygen and mixed with the crude helium into the first stage dehydrogenation reactor, which outlet temperature is controlled at about 200 °C. Thirdly, the crude helium after dehydrogenation is cooled down to 80–120 °C by a primary water cooler, and then mixed with oxygen into the second stage dehydrogenation reactor, which outlet temperature is controlled at about 200 °C. Lastly, the dehydrogenated crude helium is cooled down to 80–120 °C by a secondary water cooler, mixed with supplementary oxygen once more and introduced into the three-stage dehydrogenation reactor. After the reaction, the gas is cooled to around 40 °C through the third-stage water cooler, where free water is separated and the hydrogen content in the outlet gas is controlled to not exceed 0.5 mL/m^3. Then the dehydrogenated crude helium undergoes dehydration.

The crude helium is dewatered using molecular sieve adsorption, typically equipped with two dehydration towers (Fig. 5.3). The step-down regeneration process is applied, regeneration gas can be nitrogen or crude helium. During operation, one tower is kept adsorbing while the other performs thermal blow or cold blow. After the dehydrogenation reaction, crude helium enters the crude helium dehydration tower for adsorption and dehydration. The dehydrated crude helium then passes through a dust filter and enters downstream units for processing. Once the dehydration tower completes the adsorption process, the saturated crude helium adsorbed in the dehydration tower is released and recovered in the helium recovery gas holder.

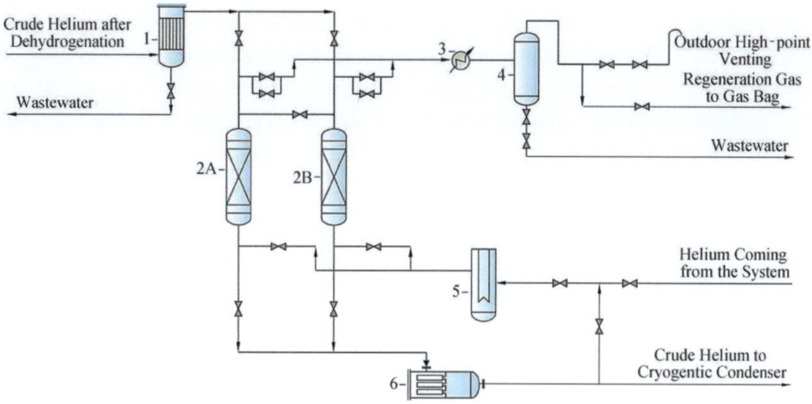

1 - Wet Crude Helium Aggregator; 2A/2B - Crude Helium Adsorption Tank; 3 - Crude Helium Regeneration Gas Cooler; 4 - Crude Helium Regeneration Gas Separator; 5 - Crude Helium Regeneration Gas Electric Heater; 6 - Dry Crude Helium Dust Filter

Fig. 5.3 Process flow of dual-column dehydration

Taking nitrogen regeneration as an example, regeneration gas is heated to 250 °C in an electric heater and regenerated in the dehydration tower. Water adsorbed by the molecular sieves is desorbed due to the high temperature of the regeneration gas and is discharged along with the regeneration gas at a high point. After completing the heating process, the electric heater stops heating. The unheated nitrogen is used as cold blow gas to cool the bed layer. Cold blow is considered complete when the outlet temperature drops to 40–50 °C.

5.2 Adsorptive Dehydrogenation Technology

Adsorptive dehydrogenation technology utilizes metal hydrides as adsorbents to absorb hydrogen, thereby achieving the separation of hydrogen from helium. As hydrogen is absorbed, the remaining helium and other components in the feed gas are separated and enriched. This technology is particularly suitable for crude helium with high hydrogen content, such as the exhaust gas produced in the ammonia synthesis industry, where hydrogen content can be as high as 50% and helium content ranges from 0.3% to 3%.

5.2.1 Principle

Metal hydride adsorption dehydrogenation leverages the selective absorption properties of hydrogen storage alloys (M) towards hydrogen. Under specific temperature and pressure conditions, these alloys react with gaseous hydrogen to produce metal hydrides (MH_x). The reaction mechanism of metal hydride adsorption dehydrogenation involves the adsorption of hydrogen molecules on the metal surface, followed by their dissociation into hydrogen atoms, which are then stored in the interstitial lattice sites of the hydride's crystal structure. The reaction formula is as follows:

$$M + \frac{x}{2}H_2 \Leftrightarrow MH_x + Q \quad (5.8)$$

where M represents the solid solution or intermetallic compound with a defined crystal lattice structure and elemental ratio; and MH_x is the corresponding hydride; x is the atomic ratio of hydrogen to the metal; Q is the reaction heat.

Typically, the internal energy of the hydride is lower than that of the metal and gaseous hydrogen. Therefore, the formation of hydrides is exothermic at room temperature and higher, while the release of hydrogen from hydride is endothermic, requiring energy input for the regeneration of the hydride.

This reaction is reversible. By increasing the system pressure or decreasing the temperature, the reaction proceeds to the right to form hydride, and the impurities gas is left in the metal lattice and does not participate in the reaction, achieving hydrogen removal. Conversely, by lowering the pressure or increasing the temperature, the metal hydride decomposes, releasing hydrogen at a purity level of up to 99.9999%. This method features simple, noiseless, and safe operation. To prevent poisoning of the metal hydride, the concentrations of sulfur, carbon monoxide, and carbon dioxide in the crude helium should be kept below 1 mL/m^3 (ppmV).

5.2.2 *Typical Process Flow*

Take a typical three-column adsorptive dehydrogenation process as an example, as shown in Fig. 5.4.

Crude helium containing hydrogen enters the metal hydride adsorption tower 1A, where hydrogen is absorbed by the metal hydride, while helium and other components proceed to the next step. Adsorption tower 1B regenerates metal hydride by heating with thermal oil at 220 °C. During regeneration, hydrogen is desorbed. Once regeneration is complete, tower 1C cools down the desorbed gas using thermal oil at 5 °C, ensuring further hydrogen desorption. Vacuumizing is employed to ensure complete regeneration of the metal hydride. After tower 1A reaches adsorption saturation, it enters a heating regeneration cycle using thermal oil to desorb hydrogen. Tower 1B enters a cooling regeneration cycle, utilizing 5 °C thermal oil for cooling, the use of vacuum to ensure that the metal hydride regeneration completely. Tower 1C enters the adsorption cycle where crude helium containing hydrogen enters, hydrogen is adsorbed, and helium passes through tower 1C to the next step.

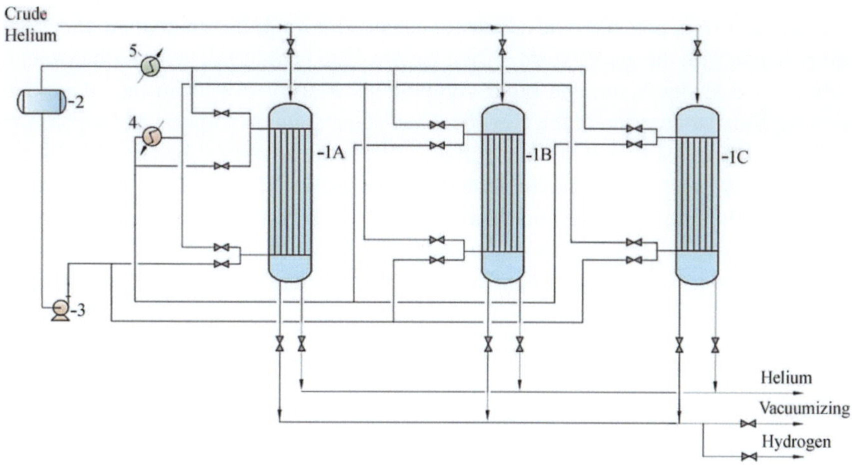

1A~1C - Dehydrogenation adsorption tower; 2 - Separator; 3 - Thermal Oil Pump; 4 - Heater; 5 - Cooler

Fig. 5.4 Process flow of three-column adsorptive dehydrogenation

5.3 Pressure Swing Adsorption Technology

The pressure swing adsorption technology can be employed to remove water, carbon dioxide, carbon monoxide, methane, nitrogen, and other impurities from crude helium for the purpose of purifying the gas. This technology is characterized by low energy consumption, low investment, and convenient operation, making it the most widely used method for gas separation.

5.3.1 Principle

The pressure swing adsorption technology utilizes the differences in polarity, molecular weight, or other chemical properties of different gas molecules, as well as the differential adsorption capacities on different adsorbents, to reach the purpose of helium purification.

Depending on the specific component to be separated, specific adsorbents are used to purify and separate the desired component. The order of adsorption capacities of common gas molecules on adsorbents is as follows: $He < H_2 < Ar < N_2 < CO < CH_4 < CO_2 < C_2H_6 < C_3H_8 < H_2S < H_2O$. In the application of helium purification, the characteristics of helium's weaker adsorption properties than other gases are utilized to select the appropriate adsorbent. Crude helium passes through the adsorbent bed under high pressure. Water, carbon dioxide, methane, nitrogen, and other components are selectively adsorbed, while helium passes through the adsorbent bed because it is not easily adsorbed, thus realizing the separation of the components.

Under a certain pressure, the adsorbent can adsorb a significant amount of certain components from the gas mixture. When the pressure is reduced, the adsorbed water, carbon dioxide, methane, and other components desorb fully from the adsorbent, allowing the regeneration of the adsorbent and subsequent adsorption and separation cycles, realizing the recycling of adsorbent. When the adsorbent bed becomes saturated, several pressure equalization steps are required. This helps in the complete recovery of helium from the non-flow region of the bed and increases the impurity concentration in the dead space of the bed.

The desorption of the adsorbent is achieved by reducing the partial pressure of the adsorbed component. Common methods include:

(1) Reduce the pressure of the adsorbent bed (pressure equalization or pressure release).
(2) Rinse with product gas components.
(3) Evacuate using a vacuum pump.

The pressure and adsorption capacity curves during the adsorption and desorption processes of the adsorbent bed are shown in Fig. 5.5.

With regard to the selection of adsorbents, Das et al. established a cascaded PSA pilot plant in 2008 to recover helium from natural gas [1]. The process employed four-stage pressure swing adsorption for stepwise purification of helium. The first-stage adsorption tower was filled with silica gel adsorbent to remove large molecules such as hydrocarbons and carbon dioxide. The remaining gases with poorer adsorption characteristics, including methane, nitrogen, and helium, entered the second-stage adsorption tower, which contained activated carbon to remove a portion of methane. The third and fourth-stage adsorption towers were filled with 13X zeolite adsorbent to further eliminate other small molecular impurities and thus enhance helium purity.

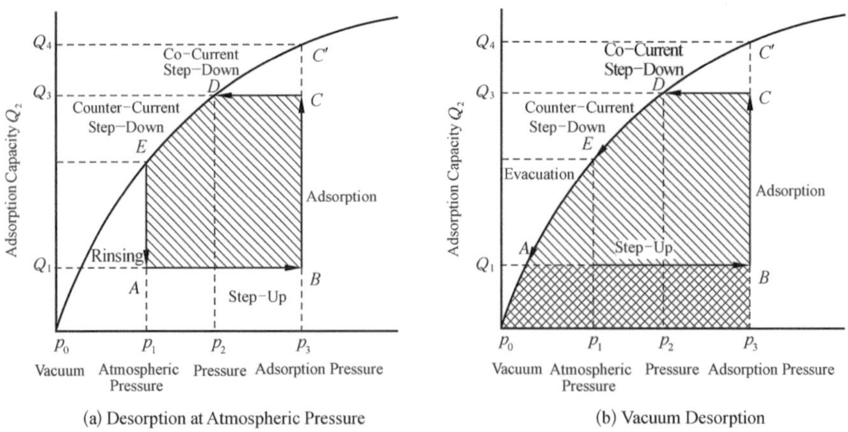

Fig. 5.5 Pressure and adsorption capacity curves during the adsorption and desorption processes of the adsorbent bed

Through this cascaded system, 99% helium can be extracted from natural gas with a helium content of 0.06%, with a helium recovery rate of 61%.

Theoretically, PSA can be used independently for helium purification. However, in practice, due to the generally low helium content of crude helium, a large amount of adsorbent is required to adsorb all other gases except helium and hydrogen from the natural gas, which can lead to a significant increase in plant investment. Furthermore, the desorption of impurity gases from crude helium through PSA at atmospheric pressure requires reducing the pressure to atmospheric pressure and then compressing it back to the original pressure, leading to high equipment investment and operational energy consumption. Therefore, solely employing PSA for helium purification struggles to yield expected results. Typically, PSA is used to purify crude helium with a certain concentration to a helium concentration of approximately 99.9%.

5.3.2 *Typical Process Flow*

Take a typical six-tower PSA adsorption process as an example, as shown in Fig. 5.6.

Each adsorption tower goes through several distinct steps in one cycle as follows: adsorption (AD), multiple equalization drops (E_nD), counter-current depressurization (D), counter-current evacuation (V), multiple equalization rises (E_nR), and final repressurization (FR). The operational sequence for the six PSA towers is provided in Table 5.1, and the definitions of the steps are given in Table 5.2.

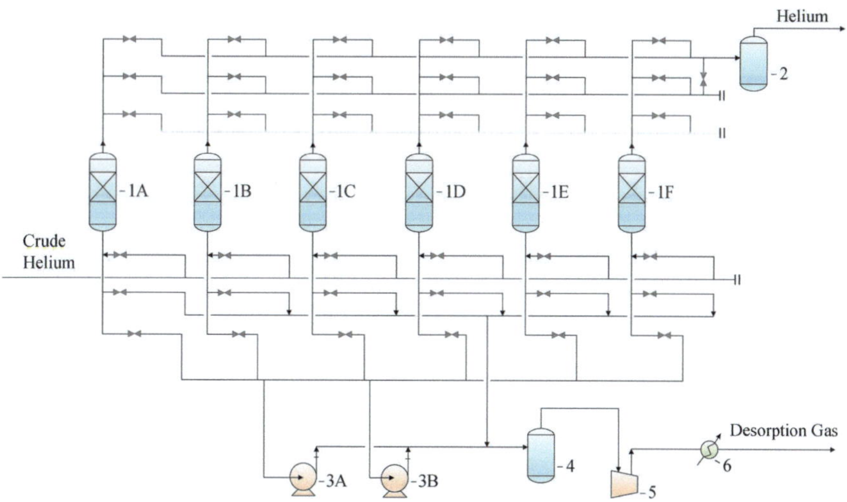

1A~1F - Adsorption Tower; 2 - Product Gas Outlet Buffer Tank; 3A/3B - Vacuum Pump; 4 - Desorption Gas Buffer Tank; 5 - Desorption Gas Compressor; 6 - Cooler

Fig. 5.6 Flow chart of six-PSA-tower adsorption

Table 5.1 Operational sequence of adsorption towers

Step no.	1	2	3	4	5	6	7	8	9	10	11	12
Step position	T1	T2	T3	T4	T1	T2	T3	T4	T1	T2	T3	T4
A	AD				E_1D	E_2D	E_3D	D	V			
B	E_1R			FR	AD				E_1D	E_2D	E_3D	D
C		E_2R			E_1R			FR	AD			
D			E_3R			E_2R			E_1R			FR
E	V						E_3R			E_2R		
F	E_1D	E_2D	E_3D	D	V						E_3R	
Step no.	13	14	15	16	17	18	19	20	21	22	23	24
Step position	T1	T2	T3	T4	T1	T2	T3	T4	T1	T2	T3	T4
A			E_3R			E_2R			E_1R			FR
B	V						E_3R			E_2R		
C	E_1D	E_2D	E_3D	D	V						E_3R	
D	AD				E_1D	E_2D	E_3D	D	V			
E	E_1R			FR	AD				E_1D	E_2D	E_3D	D
F		E_2R			E_1R			FR	AD			

Table 5.2 Definitions of adsorption tower steps

Time no.	Meaning	Preset value	Setting principle
T1 (step 1)	First equalization time	60s	Ensure first pressure equalization
T2 (step 2)	Second equalization time	60s	Ensure second pressure equalization
T3 (step 3)	Third equalization time	60s	Ensure pressure equalization
T4 (step 4)	Counter-current depressurization, final repressurization time	60s	Ensure final repressurization time, minimize pressure fluctuations in product gas

Notes Single-tower adsorption time = T1 + T2 + T3 + T4; vacuumizing time = T1 + T2 + T3 + T4; complete PSA cycle time = (T1 + T2 + T3 + T4) × 6

Taking Tower 1A as an example, the specific steps are as follows:

(1) Adsorption (A). The crude helium gas enters the pressure swing adsorption (PSA) unit through a programmed control valve at the bottom of adsorption tower 1A. As the gas flows upward through the adsorption bed under working pressure, methane and nitrogen in the gas stream are selectively adsorbed by the adsorbent. The qualified product gas is discharged from the top of the tower via the pipeline and transported outside the system. The feed is stopped and the adsorption process is terminated when the adsorption front of

impurities approaches the outlet of the adsorption bed. At this point, a section of "unadsorbed adsorbent" (reserved section) remains between the adsorption front and the outlet end.

(2) First Equalization Drop (E_1D). After the adsorption in tower 1A ends, connect the outlets of towers 1A and 1C to equalize their pressures. The adsorption front in tower 1A continues to move forward. When the pressures in towers 1A and 1C are nearly equal, the adsorption front in tower 1A has not yet reached the outlet. This process helps recover gas in the non-flowing region of tower 1A, which has essentially the same composition as the output intermediate gas.

(3) Second Equalization Drop (E_2D). Tower A is pressure-equalized with tower D, similar to the E_1D process.

(4) Third Equalization Drop (E_3D). Tower A is pressure-equalized with tower E, similar to the E_1D process.

(5) Counter-Current Depressurization (D). The effective gas is released from adsorption tower 1A through counter-current depressurization. The desorbed gas is recovered by the desorption gas compressor into a gas holder.

(6) Counter-Current Evacuation (V). To further regenerate the adsorbent, a vacuum pump is used to perform a counter-current evacuation of the adsorption tower. The bottom of tower 1A is connected to the vacuum pump inlet, and the outlet gas is vented. Evacuation ends when the pressure inside the tower reaches −0.09 MPaG.

(7) Third Equalization Rise (E_3R). After the evacuation, tower 1A is pressure-equalized with tower 1C via a sequencing valve, i.e. connecting the outlets of both towers. Tower 1C performs E3R on tower 1A.

(8) Second Equalization Rise (E_2R). Tower 1A is pressure-equalized with tower 1D, similar to the E_3R process.

(9) First Equalization Rise (E_1R). Tower 1A is pressure-equalized with tower 1E, similar to the E_3R process.

(10) Final Repressurization (FR). After equalization rise, tower 1A is finally repressurized with product gas until its pressure equals the adsorption pressure.

After final repressurization, the pressure in the adsorption tower is close to the adsorption pressure, ready for the next adsorption (AD) cycle, and can be programmed to enter the next cycle.

This series of pressure equalization processes not only recovers helium from the non-flowing regions of the towers and increases the pressure in tower 1A but also directs residual impurities towards the inlet end of the adsorption tower, thereby "refreshing" the adsorbent.

The operating steps for the rest five towers are identical to that of tower 1A, though they are staggered from each other in time according to a certain procedures.

5.4 Temperature Swing Adsorption Technology

Temperature swing adsorption (TSA) technology is known for its simple process, low investment, and straightforward operation, making it one of the earliest cyclic adsorption processes to be industrialized. It finds applications in gas drying, feed gas purification, waste gas removal, as well as exhaust gas and waste liquid treatment in environmental protection. This technology, in principle, can also effectively remove water, carbon dioxide, carbon monoxide, methane, nitrogen, and other impurities from crude helium, achieving the purpose of helium purification. In the helium purification scenario, there are patents pertaining to TSA technology, but no industrial application cases have been found.

5.4.1 Principle

TSA operates based on the characteristic of adsorbents where the equilibrium adsorption capacity decreases as temperature rises. This technology achieves helium purification through the process of adsorption at ambient temperature followed by desorption at elevated temperatures. It is worth noting that, after adsorption, the adsorbent needs to undergo drying, cooling, and other processes to complete the adsorption cycle.

The change in Gibbs free energy (ΔG) during adsorption at a constant pressure is related as follows:

$$G = \Delta H - T\Delta S \tag{5.9}$$

where: ΔH is the enthalpy change; ΔS is the entropy change.

When adsorption reaches equilibrium, both the free energy and entropy decrease. ΔH in the equation is negative, indicating that the adsorption process is exothermic. It is thus evident that lowering the operating temperature increases the adsorption capacity, and vice versa. Therefore, adsorption is typically carried out at lower temperatures, followed by heating to desorb the adsorbed components. Steam is often used for direct heating of the adsorbent to facilitate desorption, with the desorbed components separated from the steam after condensation. The adsorbent undergoes stages of indirect heating, drying, and cooling to enable its cyclic reuse.

Commonly used adsorbents in TSA processes include activated carbon, molecular sieves, activated alumina, and silica gel, among others. Selecting the appropriate adsorbent based on specific process requirements is one of keys to TSA technology.

5.4.2 *Typical Process Flow*

An example of an integrated helium extraction process using TSA and Vacuum Pressure Swing Adsorption (VPSA), as described in a U.S. patent, is illustrated in Fig. 5.7 [2].

The TSA unit primarily removes methane and heavier components from the feed to prevent rapid degradation and saturation of the VPSA unit, thereby ramping up efficiency and recovery rates of the VPSA unit.

5.5 Low-Temperature Adsorption Technology

Low-temperature adsorption technology is effective for removing trace impurities such as argon, nitrogen, and neon from crude helium to achieve the purpose of helium purification. This technology is well-developed and widely used, yet it consumes a large amount of liquid nitrogen.

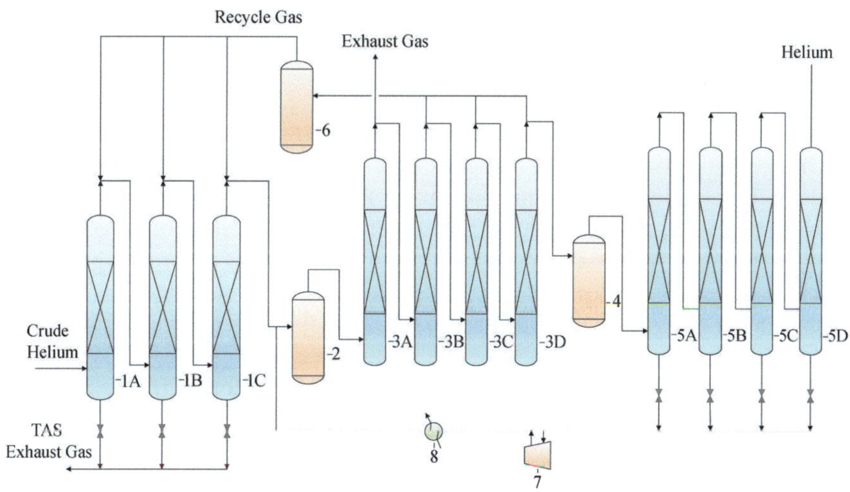

1A~1C - TSA Adsorption Tower Unit; 2 - Buffer Tank; 3A~3D - First-Stage VPSA Adsorption Tower Unit; 4 - Buffer Tank; 5A~5D - Second-Stage VPSA Adsorption Tower Unit; 6 - Buffer Tank; 7 - Compressor; 8 - Cooler

Fig. 5.7 TSA + vacuum PSA process flow

5.5.1 Principle

Low-temperature adsorption Technologies generally operates at liquid nitrogen temperatures and leverages the adsorption characteristics of adsorbents in a low-temperature environment to absorb residual impurity components, elevating helium purity to 99.999%.

After undergoing dehydration and dehydrogenation, the crude helium is cooled in the cold box of a low-temperature adsorption unit and then enters a separator. The uncondensed gas phase is subjected to nitrogen removal through adsorption using a nitrogen adsorber in the nitrogen removal cold box to remove the remaining nitrogen from crude helium. At this stage, neon remains in the helium. Neon-containing helium then exits the nitrogen removal cold box and returns to the absorptive nitrogen removal model, where it is reheated to ambient temperature via a heat exchanger, and then pressurized by a diaphragm compressor before entering the neon removal system. The neon component in the crude helium is adsorbed and removed in the neon removal system, resulting in product-grade helium [3].

5.5.2 Typical Process Flow

Take a typical cryogenic adsorption process flow using two sets of adsorbers as an example, as depicted in Fig. 5.8.

The operating principle of the cryogenic adsorption unit involves low-temperature condensation and low-temperature adsorption for separation and purification. The typical process sets up two switching adsorbers, each consisting of an adsorption cylinder immersed in liquid nitrogen. The adsorbers utilize the adsorption properties of activated carbon in a low-temperature environment to remove residual impurity components, raising the helium content to 99.999%. The regeneration of the adsorbers adopts heating regeneration, when the adsorption cylinder as a whole is reheated to room temperature, the basic desorption of impurities adsorbed by the activated carbon is completed. The detailed steps are as follows:

(1) Drying Process

Crude helium enters a dryer to ensure that the gas entering the cryogenic adsorbers is dry, preventing any potential freezing and blockage issues.

(2) Heat Exchange Process

Since the cryogenic adsorption process operates at liquid nitrogen temperatures, to minimize the loss of cooling capacity (liquid nitrogen) during purification, there should be heat exchanges between the unpurified incoming gas and the purified outgoing gas in a high-pressure bushing heat exchanger.

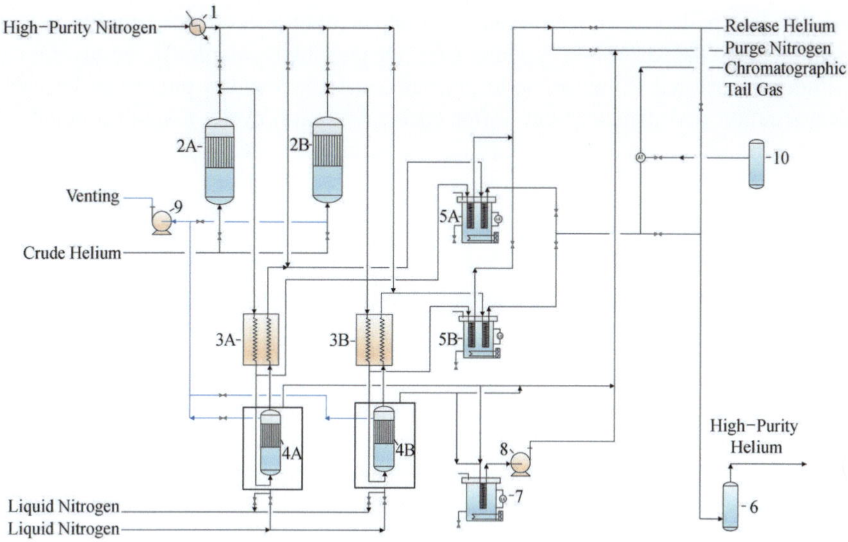

1 - Nitrogen Heater; 2A/2B - Dryer; 3A/3B - Heat Exchanger; 4A/4B - Adsorber; 5A/5B - Helium Water Bath Vaporizer; 6 - Gas Supply Buffer Tank; 7 - Nitrogen Water Bath Vaporizer; 8 - Evacuation Pump; 9 - Activation Vacuum Pump; 10 - Helium Cylinder

Fig. 5.8 Process flow of cryogenic adsorption

(3) Switching and Purification Process

Each set of adsorbers is pre-filled with liquid nitrogen into the nitrogen storage Dewar five hours before the scheduled switch time. The middle part of the adsorber's Dewar is equipped with a differential pressure level gauge, which is used to control the opening and closing of the liquid nitrogen valve. When the switch time is reached, both sets of adsorbers operate simultaneously. Subsequently, the inlet and outlet valves of the saturated adsorber are closed, and the activation process begins. The standby adsorber enters the automatic purification process.

During the purification operation of the adsorber, when the purity of helium is detected to be less than 99.999%, the impure helium valve is opened to return to the front end for recycling and purification. Once the purity of helium at the outlet of the adsorber reaches above 99.999% and stabilizes, the product gas valve is opened to enter downstream storage facilities.

If negative pressure liquid nitrogen is required, the vacuum degree inside the Dewar can be controlled by adjusting the opening of the proportion valve at the inlet of the vacuum pump, to control liquid nitrogen Dewar vacuum degree stabilization.

(4) Activation Regeneration Process

The reactivation of the adsorbers adopts the methods of purging with hot nitrogen gas and evacuating (or applying vacuum). Before activation regeneration, residual helium is discharged to the pressure relief pipeline. Once pressure relief is completed, liquid nitrogen is expelled, and hot nitrogen purge and heating coils operate simultaneously to rapidly heat the adsorption cylinder. After the purging is completed, the adsorption cylinder is restored to atmospheric pressure, and the vacuum pump can be used to evacuation and displacement. After each evacuation cycle, a small amount of high-purity helium is used to backfill the adsorption cylinder to get a slight positive pressure.

5.6 Low-Temperature Pressure Swing Adsorption Technology

Low-temperature pressure swing adsorption technology (LTPSA) is employed for the purification of crude helium by removing trace impurities such as argon, nitrogen, and neon. This technique is relatively novel and offers the advantage of regenerating adsorbers at low temperatures, significantly reducing the consumption of liquid nitrogen and showing a certain potential for industrial application.

5.6.1 Principle

LTPSA technology purifies and refines helium by leveraging the principles of PSA at low temperatures. While it shares the core principle with low-temperature adsorption—using an adsorbent to remove impurities—the key difference lies in the use of pressure variation to adopt desorption under low temperatures.

This technology typically operates at liquid nitrogen temperatures, around 66 K (− 207 °C). The feed gas enters the bottom of the adsorber, where impurities such as neon are adsorbed by the adsorbent. Product helium is then collected from the top of the adsorber. Once the adsorbent becomes saturated, it undergoes a desorption process involving counter-current depressurization, evacuation, and rinsing to regenerate the adsorbent. The desorbed gas is pressurized and recirculated back into the previous processing stage for reuse. Each adsorber undergoes a sequence of adsorption, co-current depressurization, equalization drop, counter-current depressurization, evacuation and rinsing, equalization rise, and final repressurization. The steps for each adsorber are staggered to ensure continuous and stable operation of the entire system.

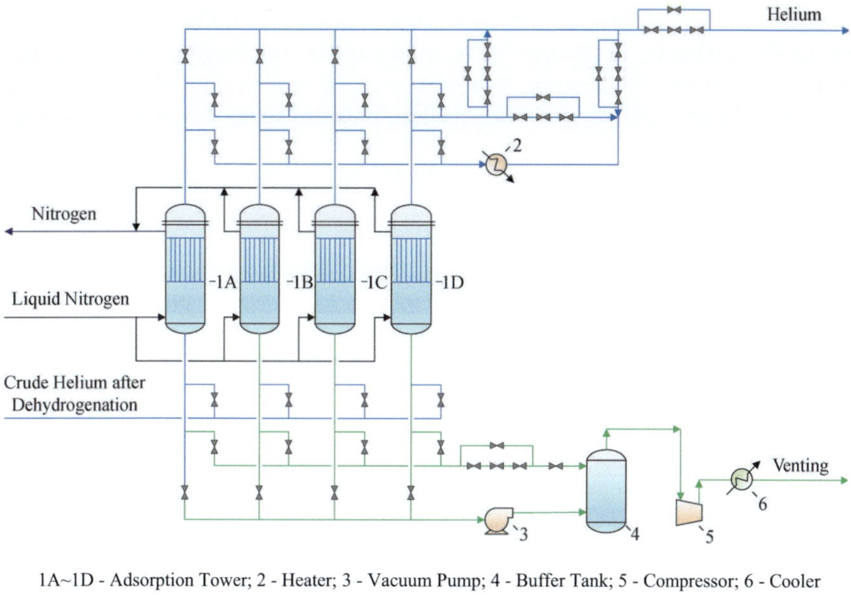

Fig. 5.9 LTPSA process flow

5.6.2 *Typical Process Flow*

The Southwest Research and Design Institute of the Chemical Industry has carried out relevant technical research and developed a novel low-temperature refining technology. This technology has been successfully implemented on an industrial scale in November 2022. The process flow is illustrated in Fig. 5.9.

This technology offers several advantages:

(1) Efficient and streamlined process, with simple operation and a small footprint.
(2) Significantly shortened adsorption cycles, with each cycle lasting only 0.5–2 h.
(3) Elimination of frequent temperature fluctuations during regeneration, which reduces energy consumption and extends the service life of both the adsorbent and the adsorption unit.

5.7 Technical Applications

Each of the aforementioned technologies has its own characteristics, and in practical engineering applications, it is suggested to select a suitable combination of processes based on the composition and content of crude helium, while taking into account factors such as economic feasibility and operability.

Taking a specific project as an example, there is a crude helium refining unit designed to remove impurities from crude helium and obtain high-purity helium products. The overall process flow is illustrated in Fig. 5.10. The process involves membrane separation technology in the front end for enriching the helium content from 0.14% to 63.45%, and the back-end crude helium refining utilizes a typical scheme covering “catalytic dehydrogenation, pressure swing adsorption, and low-temperature adsorption”.

1. Catalytic Dehydrogenation Unit

The catalytic dehydrogenation unit utilizes a three-stage catalytic oxidation scheme with a designed processing capacity of 675 m^3/d (at 20 °C and 101.325 kPa, unless otherwise specified). After dehydrogenation, the crude helium contains less than 600 mL/m^3 (ppmV) of O_2, less than 0.5 mL/m^3 (ppmV) of H_2, and less than 1 mL/m^3 (ppmV) of H_2O.

Inside the dehydrogenation unit, hydrogen content is monitored at the inlet using an online hydrogen analyzer, and hydrogen and oxygen analyzers are installed at the outlet to ensure the final dehydrogenation passes muster. After dehydrogenation, crude helium undergoes water removal using molecular sieve adsorption. Oxygen is supplied from oxygen cylinders and introduced into the unit after pressure relief. The unit features a built-in circulating chilled water system. The initial conditions of the feed gas for the catalytic dehydrogenation unit are as follows: feed gas pressure of 1.2 MPa, temperature of 40 °C, and flow rate of 608 m^3/d.

The composition of the feed gas and product gas from the catalytic dehydrogenation unit is detailed in Table 5.3.

2. PSA Unit

A PSA unit is employed to remove impurities such as carbon dioxide, methane, water, and oxygen from crude helium, thereby producing a helium purity of over 99.9%. Given the relatively high content of carbon dioxide in crude helium, a two-stage PSA scheme is utilized for refining. The unit is designed with a processing capacity of 675 m^3/d and equipped with an online chromatographic analyzer at the outlet to ensure the purity of crude helium. Non-compliant gas is automatically switched and vented to the second-stage PSA desorption gas buffer tank, then compressed and returned to the membrane separation unit.

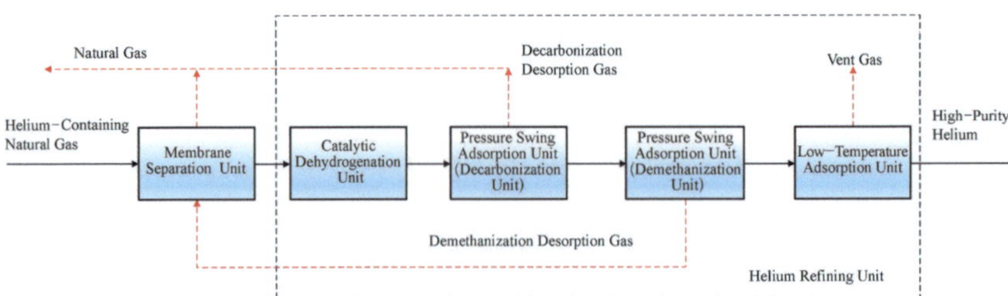

Fig. 5.10 Process flow of a helium extraction project

Table 5.3 Composition of feed gas and product gas from the catalytic dehydrogenation unit

Component	Feed gas/% (mole fraction)	Product gas/% (mole fraction)
Methane	24.8076	25.3964
Ethane	0.0274	0.0376
Propane	0.0009	0.0011
Carbon dioxide	4.8470	4.9621
Nitrogen	4.4778	4.5841
Helium	63.4539	64.9602
Hydrogen	2.3805	0.0000
Oxygen	0.0020	0.0585
Water	0.0029	0.0000
Total	100.0000	100.0000

The first-stage PSA unit comprises six adsorption towers earmarked to remove carbon dioxide and partial methane and nitrogen. The forward-flowing gas from this unit is mixed with desorption gas from the second stage and pressurized by the decarbonization desorption gas compressor to a pressure of 3.5 MPa for delivery to the external pipeline network.

The second-stage PSA unit also consists of six adsorption towers. The forward-flowing gas from this unit is mixed with the desorption gas from the first and second stages, which is primarily composed of methane, nitrogen, and helium. The mixture gas is pressurized by a demethanization desorption gas compressor to 3.5 MPa and fed back to the second-stage membrane inlet to boost helium recovery. The initial conditions of the feed gas for the PSA unit are: feed gas pressure of 1.14 MPa, temperature of 40 °C, and flow rate of 600 m^3/d.

The composition of the feed gas and product gas from the PSA unit is detailed in Table 5.4.

Table 5.4 Composition of feed gas and product gas from the PSA unit

Component	Feed gas/% (mole fraction)	Product gas/% (mole fraction)	Decarbonization desorption gas/% (mole fraction)	Demethanization desorption gas/% (mole fraction)
Methane	25.3964	0.0000	64.9918	39.2853
Ethane	0.0376	0.0000	0.2368	0.0000
Propane	0.0011	0.0000	0.0069	0.0000
Carbon dioxide	4.9621	0.0000	31.2307	0.0000
Nitrogen	4.5841	$\leq$ 0.1000	0.4948	11.3181
Helium	64.9602	$\geq$ 99.9000	2.8958	49.3099
Oxygen	0.0585	0.0000	0.1432	0.0867
Total	100.0000	100.0000	100.0000	100.0000

Table 5.5 Composition of feed gas and product gases from the low-temperature adsorption unit

Component	Feed gas/% (mole fraction)	Product gas/% (mole fraction)
Helium	$\geq$ 99.9000	$\geq$ 99.9990
Impurity	$\leq$ 0.1000	$\leq$ 0.0010
Total	100.0000	100.0000

3. Low-Temperature Adsorption Unit

A low-temperature adsorption unit is installed to further remove impurities such as nitrogen, water, oxygen, neon, and others from crude helium, thereby achieving a high-purity helium product of 99.999%. The single-column unit is designed with a processing capacity of 300 m^3/d.

For handling non-compliant and regeneration gases, a small amount of helium regenerated from the low-temperature adsorption unit is reheated and sent to the helium recovery gas holder for introduction into the desorption gas buffer tank of the PSA unit, where it is recycled along with desorption gas. The product gas outlet is equipped with one online chromatographic analyzer.

The unit includes a 30 m^3 liquid nitrogen storage tank and associated vaporizer. The liquid nitrogen in the liquid nitrogen storage tank provides cooling for the cryogenic adsorption skid and purging for regeneration of the cryogenic adsorption skid. Initial conditions of the feed gas for the low-temperature adsorption unit are: feed gas pressure of 1.05 MPa, temperature of 40 °C, and flow rate of 272 m^3/d.

The composition of the feed gas and product gas from the low-temperature adsorption unit is detailed in Table 5.5.

References

1. Das NK, Chaudhuri H, Bhandari RK, et al. Purification of helium from natural gas by pressure swing adsorption. Curr Sci. 2008;95(12):1684–1687.
2. Joseph SD, Herbert ER, Kent SK. Helium recovery: US 5542966. 1996-08-06.
3. Taiyo Sanso Co Ltd. Helium-gas liquefaction apparatus has cooling head of cryogenic refrigerator inserted into vacuum insulating tank such that helium gas in helium-gas pipeline is cooled below liquefying temperature: JP 2005083588-A. 2005-3-31.

Chapter 6
Helium Liquefaction Technologies

Miao Yang and Limin Qiu

Temperature is a physical quantity that measures the degree of hotness or coldness of a substance, which, on a microscopic level, represents the intensity of the thermal motion of its molecules. At high temperatures, the theoretical upper limit is when the heaviest microscopic particles move at the speed of light, known as the Planck Temperature, with a value reaching 10^{32} K. On the other end of the spectrum lies the ultimate low temperature, termed "absolute zero", corresponding to 0 K. When converted to the commonly used temperature unit, this is equivalent to − 273.15 °C, a theoretical limit that cannot be physically attained.

The liquefaction of helium hinges on breakthroughs in cryogenic technology, with helium's liquefaction point at 4.22 K (− 268.93 °C), just about 4 degrees above "absolute zero". Over the past century, cryogenics has witnessed remarkable progress. Techniques such as pressure-reduction cooling, ^{3}He refrigeration, adiabatic demagnetization, and dilution refrigeration have evolved and become commercialized, progressively lowering achievable temperatures from 4.22 to 1 K, 500 mK, 100 mK, and eventually below 10 mK. An ambitious goal cannot be achieved in one step, as the "absolute zero" challenge has become an epic quest to liquefy gases. Thus, before delving into the specifics of helium liquefaction technology, this chapter first traces the historical journey toward these temperature extremes.

M. Yang (✉)
China Petroleum Engineering & Construction Corporation Southwest Company, Chengdu, China
e-mail: yangmiao_sw@cnpc.com.cn

L. Qiu
Institute of Refrigeration and Cryogenics, Zhejiang University, Hangzhou, China
e-mail: limin.qiu@zju.edu.cn

L. Pu (ed.), *Helium Engineering Technologies*,
https://doi.org/10.1007/978-981-96-4086-7_6

6.1 The Pursuit of "Absolute Zero"

Since 1908, the tireless efforts of scientists worldwide have propelled the technology of gas liquefaction forward. The path began with the liquefaction of chlorine, followed by oxygen, nitrogen, and subsequently hydrogen, culminating in the liquefaction of helium. Conquering each type of gas represents scaling a new summit in the field, yet the ultimate target remains the impossible and theoretical limit of "Absolute Zero". The journey of various nations and their scientists in the pursuit of "Absolute Zero" is depicted in Fig. 6.1.

6.1.1 Chlorine Liquefaction

Michael Faraday, a renowned British physicist and chemist, is celebrated as a self-taught scientist (Fig. 6.2).

On October 17, 1831, Michael Faraday made his groundbreaking discovery of electromagnetic induction, laying the foundation for the generation of alternating current. Just days later, on October 28, 1831, he invented the disc generator, the first electrical generator ever created, which changed human civilization. Due to his monumental contributions to electromagnetism, Michael Faraday is often hailed

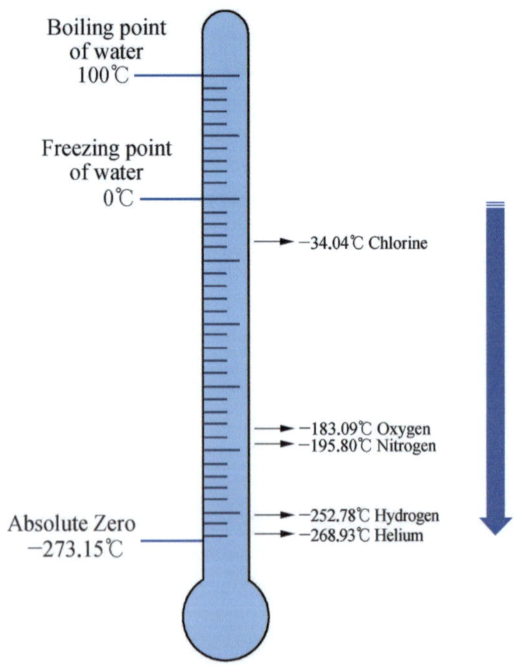

Fig. 6.1 The pursuit of "absolute zero"

Fig. 6.2 Michael Faraday. *Source* ChemTeam: Photo Gallery Five

as the "Father of Electricity" and the "Father of Alternating Current". However, Michael Faraday's contributions extend beyond electromagnetism to the field of gas liquefaction.

In 1813, Michael Faraday became the research assistant to Humphry Davy, a prominent British chemist. One of Michael Faraday's early research subjects was chlorine. Chlorine was first discovered by Swedish chemist Carl Wilhelm Scheele in 1744, and chlorine hydrate (Cl_2H_2O) was identified in 1811.

In the winter of 1823, Michael Faraday heated chlorine hydrate sealed in a glass tube. Upon cooling, he observed the formation of a colored, oily liquid. Michael Faraday found that when he heated one end of a glass tube while the other end was sealed, it was easier to obtain this liquid. Subsequent experiments revealed that this liquid was indeed chlorine, but in its liquid state. Michael Faraday concluded that this transformation was the result of low temperature and high pressure, marking the first documented instance of chlorine gas being liquefied.

Michael Faraday's discovery can be easily understood through the lens of modern science. Chlorine gas has a liquefaction temperature of − 34.04 °C at one standard atmospheric pressure. At that time, achieving such low temperatures was impossible.

However, according to thermodynamics, increasing the pressure on a gas raises its liquefaction temperature. In the sealed glass tube, the localized pressure created by the decomposition of hydrated chlorine elevated the liquefaction temperature of chlorine to a level attainable in Faraday's laboratory.

Michael Faraday's serendipitous achievement of gas liquefaction was a groundbreaking development. It marked the first time gas were artificially liquefied in a laboratory, significantly influencing the progression of science.

By 1845, using primitive compression and ice bath techniques, Michael Faraday succeeded in liquefying various gases and achieving temperatures as low as – 130 °C, the lowest possible with the technology of the time. Nevertheless, some gases, such as hydrogen, nitrogen, oxygen, and carbon monoxide, could not be liquefied even under pressures as high as 400 atmospheres. These gases were then known as "Permanent gas" [1].

6.1.2 Oxygen and Nitrogen Liquefaction

In the late 1870s, French scientist Louis Paul Cailletet became the first to successfully liquefy oxygen and nitrogen, achieving temperatures of – 183.09 °C and – 195.80 °C, respectively. His work was based on an important principle known as the Joule–Thomson Effect. Discovered by James Prescott Joule and William Thomson, the Joule–Thomson Effect refers to the temperature change observed when a gas expands through a porous plug. The process of a gas expanding through a porous plug or a throttle valve is known as adiabatic throttling expansion. This expansion is an irreversible process in which the work done by the surroundings equals the change in the internal energy of the system.

The Joule–Thomson experiment demonstrates that adiabatic throttling expansion is a process with constant enthalpy, which is a critical thermodynamic characteristic of throttling processes. However, it does not imply that the adiabatic process is always an isenthalpic process, as the intermediate states involved are non-equilibrium states. A more precise description is that gas expansion under isenthalpic conditions can lead to either an increase or decrease in temperature.

Additionally, there exists a so-called "inversion temperature". When the ambient temperature is below the inversion temperature, the gas temperature typically decreases during expansion. Conversely, if the ambient temperature is above the inversion temperature, the gas temperature increases. During isenthalpic expansion, two changes can occur: the average distance between molecules increases, leading to an increase in potential energy and a decrease in kinetic energy, which lowers the temperature; or the average distance between molecules increases, thereby reducing the average collision frequency per unit time, decreasing the potential energy converted during collisions, and increasing kinetic energy, which raises the temperature. The inversion temperatures of common gases are illustrated in Fig. 6.3.

Since the inversion temperatures of most gases are higher than room temperature, their temperature typically drops during expansion. However, hydrogen and helium

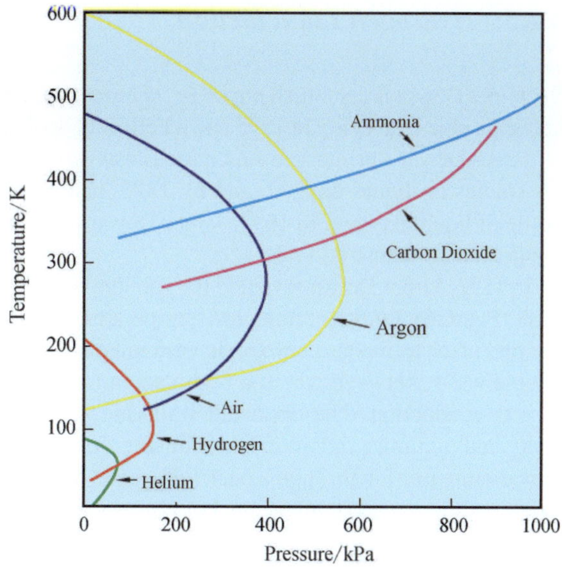

Fig. 6.3 Inversion temperatures of common gases [2]

(which had not been separated at the time) are exceptions, as their inversion temperature are much lower than the ambient temperature, causing their temperature to increase upon expansion.

Using gas adiabatic throttling processes, temperatures as low as approximately 150 K can be achieved, as shown in Fig. 6.4. For gases such as hydrogen, neon, and helium, which have inversion temperatures below 200 K (Fig. 6.3), the temperature actually rises after throttling, rendering their liquefaction impossible [2].

Fig. 6.4 Gas liquefaction apparatus from 1877 [3]

6.1.3 Hydrogen Liquefaction

Sir James Dewar is a Scottish physicist, chemist, and inventor. He became a professor at Cambridge University in 1875 and a Fellow of the Royal Society in 1877 (Fig. 6.5).

James Dewar's primary research focus was on cryogenics. Around 1880, he began his studies on liquid oxygen, and by 1885, had refined his techniques to collect a bottle of liquid oxygen. In 1891, James Dewar proposed a method to produce liquid oxygen on an industrial scale.

In 1892, James Dewar invented the vacuum flask, commonly known as the Dewar flask (Fig. 6.6), for researching low-temperature phenomena. This invention became the precursor to modern cryogenic containers. The Dewar flask is a container with double walls, between which a high vacuum is maintained to reduce heat transfer by gas conduction. The opposing walls are silvered or polished to reduce emissivity and minimize radiation heat transfer. The inner wall is connected to the room temperature portion through a neck tube made of low thermal conductivity materials (such as thin-walled stainless steel or non-metallic materials), thereby reducing heat

Fig. 6.5 James Dewar [4]

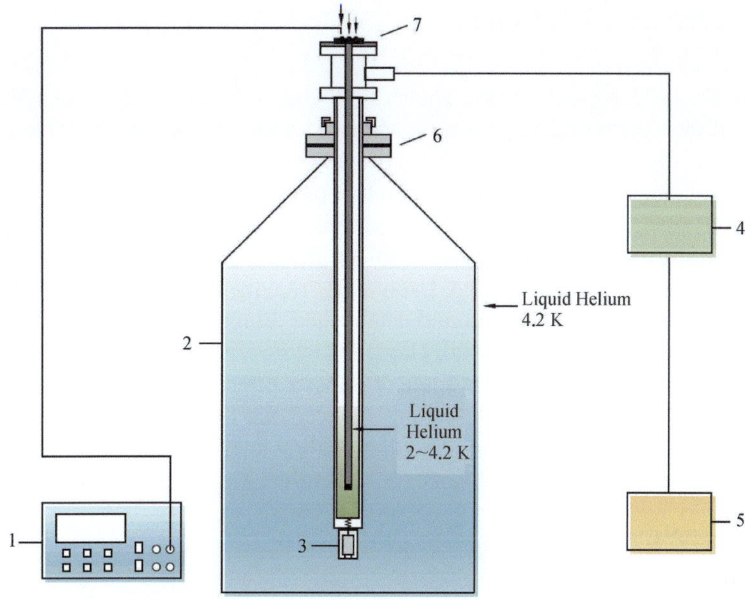

1 - Control Unit; 2 - Dewar Flask; 3 - Filter; 4 - Pressure Regulating System; 5 - Oil Pump; 6 - Dewar Flask Neck; 7 - Pressure and Temperature Connector, Optical Fiber Connector

Fig. 6.6 Schematic diagram of liquid hydrogen Dewar Flask

conduction from the room temperature portion through the neck. As thermal insulation technology advances, Dewar flasks were further improved with multilayer and multiscreen insulation to reduce heat leakage even more [5].

In 1898, James Dewar successfully liquefied hydrogen using his Dewar flask, achieving a temperature of 20.37 K. By 1899, he had also solidified hydrogen by evacuating the vapor above the solid hydrogen, reaching temperatures as low as 12 K. This work laid the groundwork for future research in cryogenics. James Dewar invented a container for storing low-temperature liquefied gases, known as a double-walled glass container with a silvered inner layer and a vacuum seal. This design was later adapted into the well-known household item, the thermos. Similarly, laboratories also required vacuum-insulated containers for the transport and storage of liquid gases. In 1906, James Dewar invented a metal Dewar flask for storing liquid oxygen, with a design capacity of 1.1×10^5 L for railway transport. The daily evaporation rate of liquid oxygen was approximately 0.1%, while for liquid hydrogen it was around 0.8%. James Dewar's ambition was to liquefy the last of the so-called "Permanent Gas"—helium. Unfortunately, helium was extremely rare, and he never had enough of it for his experiments. Although James Dewar did not achieve his goal, his extensive research on hydrogen liquefaction laid a solid foundation for the eventual liquefaction of helium.

6.1.4 Helium Liquefaction

On July 10, 1908, physicist Heike Kamerlingh Onnes (Fig. 6.7) liquefied helium—the last of the "Permanent Gas"—at the Physics Laboratory of Leiden University in the Netherlands for the first time. This milestone marked the beginning of a new era in cryogenic engineering and low-temperature physics [6].

Heike Kamerlingh Onnes' laboratory played a decisive role in low-temperature research worldwide, setting a standard for twentieth-century research institutions. Under his leadership, the Leiden Laboratory rapidly advanced James Dewar's stepwise refrigeration techniques and established a large-scale liquefaction plant. Heike Kamerlingh Onnes successfully liquefied hydrogen using the Linde–Hampson Cycle, cryogenic dewars, and the Joule–Thomson Effect, pushing the temperature limit down to − 268.93 °C. He further reduced the temperature limit to 1.5 K (about − 272 °C) using pressure-reduction cooling techniques. This extraordinary achievement earned him the nickname "Mr. Absolute Zero".

Column 6.1: The First Helium Liquefaction Process at the Leiden Laboratory

On July 9, 1908, the Leiden Laboratory produced at least 75 L of liquefied air in preparation for the final "attack". At 5:45 a.m. on July 10, the process began with the liquefaction of hydrogen, producing 20 L of liquid hydrogen to serve as a cooling source for the helium liquefaction in the afternoon. The precooling of the helium liquefier with liquid hydrogen required extreme caution. Any slight oversight during the complex preparatory operations could allow trace amounts of air to leak in, potentially solidify in the liquid hydrogen, and freeze

Fig. 6.7 Heike Kamerlingh Onnes [7]

onto the walls of the glass helium container, thereby obstructing observation and jeopardizing the entire experiment. The helium liquefaction cycle started at 4:20 p.m., at which point the core of the liquefier (the cryostat) entered a record-breaking low-temperature zone. The cryostat contained a helium gas thermometer to indicate the experimental conditions, which served as the sole guide throughout the process [6].

By 6:30 p.m., the temperature had dropped below that of liquid hydrogen, ultimately reaching 6 K (approximately – 267 °C). Heike Kamerlingh Onnes, connecting the last bottle of liquid hydrogen to the apparatus, observed no immediate change. By 7:30 p.m., the temperature stabilized around 4.22 K (– 268.93 °C). A colleague who checked on the experiment remarked that the thermometer seemed to be "standing" in a basin of liquid. Upon closer observation, Heike Kamerlingh Onnes discerned the liquid surface of helium! The experiment produced only a small amount of liquid helium, to be precise, about 60 ml.

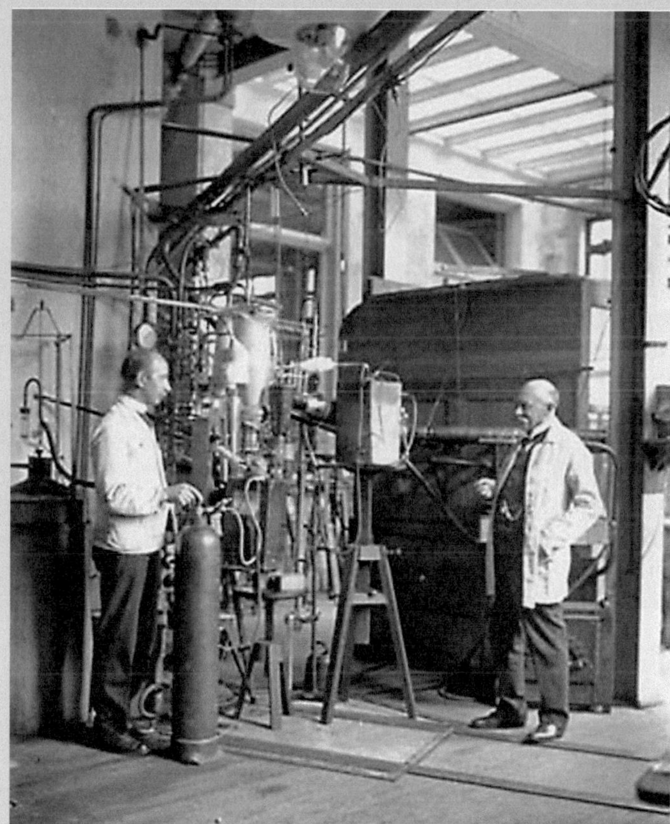

Heike Kamerlingh Onnes (right) and his chief technician, Gemt Flim, in front of the helium liquefier at the Leiden Laboratory. *Source* The super century, Nature Mater 10, 253 (2011)

In this first attempt, the Leiden Laboratory produced about 60 ml of liquid helium. In the final stage of the experiment, Heike Kamerlingh Onnes aimed to solidify helium by reducing the vapor pressure above the liquid. To achieve the lowest possible temperature, he first allowed the liquid helium to evaporate until only about 10 ml remained. He then connected the cryostat to a powerful vacuum pump, reducing the vapor pressure above the liquid to 0.01 atmospheres, yet no solid helium appeared. Consequently, Heike Kamerlingh Onnes concluded that helium's triple point must be below the lowest temperature achieved in his experiment. Unbeknownst to him, he had reached a temperature just 1 °C above "absolute zero" [6].

In 1908, Onnes' research on the properties of matter at low temperatures, particularly the production of liquid helium, earned him the 1913 Nobel Prize in Physics. In 1911, using liquid helium to cool mercury, Heike Kamerlingh Onnes discovered superconductivity, introducing the term to describe the sudden disappearance of electrical resistance in certain conductors near "Absolute Zero". This discovery opened the door to the science and technology of superconductivity [6]. Heike Kamerlingh Onnes pioneered the study of the liquid helium temperature range and superconductivity. In 1926, Willem Hendrik Keesom at Leiden Laboratory first solidified helium.

The workflow of the first helium liquefier at Leiden Laboratory is illustrated in Fig. 6.8, employing a high-pressure throttling refrigeration cycle. Here, 7 represents the compressor, and 1, 2, and 3 are three counterflow heat exchangers at different temperature stages. The high-pressure helium gas is cooled in heat exchanger 1 and further cooled by liquid nitrogen in heat exchanger 2, followed by additional cooling by liquid hydrogen in heat exchanger 3. After throttling and achieving isenthalpic expansion, the helium enters the gas-liquid two-phase region for separation, resulting in liquid helium. The remaining helium exits through the low-pressure side of the counterflow heat exchangers, exchanges heat with the high-pressure helium, and returns to the compressor inlet to continue the cycle. The liquid helium production rate is 250 milliliters per hour.

6.2 Typical Helium Liquefaction Processes

The refrigeration cycles used in helium liquefaction and refrigeration processes fall into throttling refrigeration cycle, expansion refrigeration cycle, and refrigeration cycle with gas regenerative refrigerators used for precooling, among others [8]. Each has distinct characteristics:

Fig. 6.8 Workflow of the helium liquefier

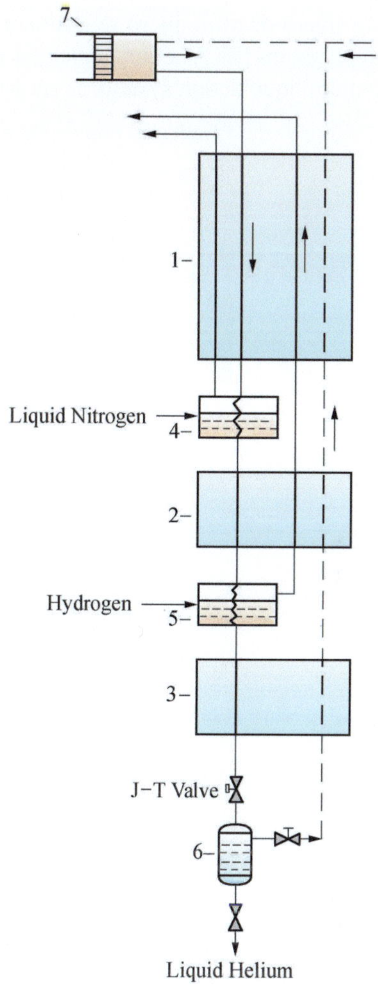

1-3 - Counterflow Heat Exchangers; 4 - Liquid Nitrogen Tank; 5 Liquid Hydrogen Tank; 6 - Gas-Liquid Separator; 7 - Compressor

6.2.1 Throttling Refrigeration Cycle

The throttling refrigeration cycle can be divided into precooling with liquid hydrogen and precooling with both liquid hydrogen and liquid nitrogen. This cycle features no moving parts and no sources of vibration, making it highly reliable. However, it relies on flammable and explosive liquid hydrogen as a pre-coolant, posing safety concerns. Additionally, due to the low efficiency of the throttling process, the energy consumption per unit of liquid hydrogen is high, making this cycle suitable only for micro or ultra-micro refrigerators, which are required for low-power, compact applications [8]. A helium liquefaction device using this cycle is known as a throttling helium liquefaction system, as shown in Fig. 6.9.

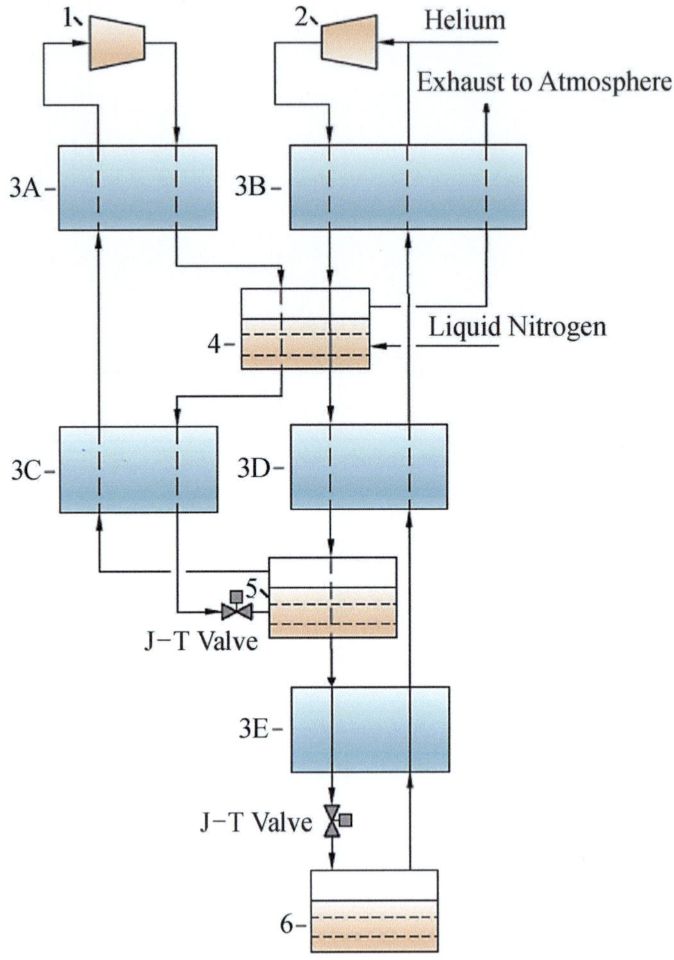

Fig. 6.9 Throttling refrigeration cycle with liquid hydrogen (or liquid nitrogen + liquid hydrogen) precooling [9]

6.2.2 Expansion Refrigeration Cycle

The expansion refrigeration cycle can operate with or without liquid nitrogen precooling. When employing multi-stage expansion refrigeration, no liquid nitrogen precooling is required, eliminating the need for any external refrigerants. However, incorporating liquid nitrogen precooling can simplify the structure by reducing the number of first-stage expanders. Adding liquid nitrogen precooling to a second-stage expander refrigeration cycle can approximately double the liquefaction capacity [8], as shown in Fig. 6.10.

The expansion refrigeration cycle offers a wide range of cooling capacity adjustment and can adapt to variable load conditions, with relatively high thermodynamic efficiency. Nonetheless, the use of expanders can affect the operational reliability of the system. With improvement of expander structure, the reliability of these systems

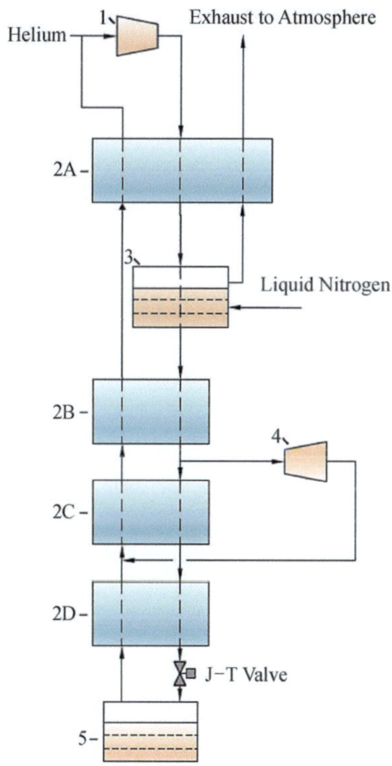

1 - Helium Compressor; 2A~2D - Heat Exchangers; 3 - Liquid Nitrogen Tank; 4 - Expander; 5 - Liquid Helium Tank

Fig. 6.10 Expansion refrigeration cycle with or without liquid nitrogen precooling [9]

has been raised, making them suitable for liquefaction systems of various capacities. A helium liquefaction device using this cycle is known as an expander helium liquefaction system.

6.2.3 Refrigeration Cycle with Gas Regenerative Refrigerator Precooling

The refrigeration cycles using gas regenerative refrigerators for precooling are compact, with a high degree of flexibility in liquid helium usage. However, they have a low liquefaction rate and long operational cycles, making them suitable for small-scale applications requiring 10–50 L per day, such as G-M refrigerators. A helium liquefaction device using this cycle is known as an alternating-flow regenerative helium liquefaction system, as shown in Fig. 6.11.

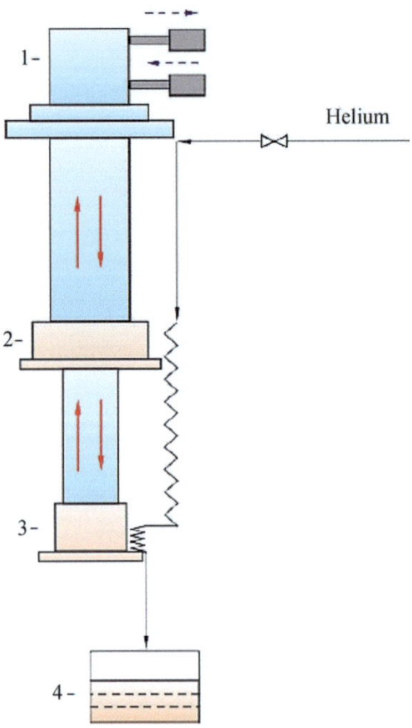

1 - GM Refrigerator; 2 - First-Stage Cold End; 3 - Second-Stage Cold End; 4 - Liquid Helium Tank

Fig. 6.11 Refrigeration cycle with gas regenerative refrigerator precooling

6.3 Typical Helium Liquefaction Process Flows

The helium liquefaction process is critical in determining the energy consumption level for liquefying helium. The process is complex and varies in form, involving steps such as compression, expansion, cooling, liquefaction, separation, and storage, each fulfilling specific functions such as pressure increase, refrigeration through expansion, and staged cooling.

(1) Compression: Helium is compressed through multiple stages to achieve a high-pressure state.
(2) Expansion: High-pressure helium is expanded through a throttle valve or a turboexpander, resulting in isothermal or adiabatic expansion to lower its temperature and pressure.
(3) Cooling: The expanded helium is further cooled via heat exchangers to reach liquefaction temperatures.
(4) Liquefaction: The cooled helium is introduced into a liquefier to convert it into a liquid state.
(5) Separation: Liquid helium is separated from non-liquefied gases to remove trace impurities, thereby ensuring the purity of the liquid helium.
(6) Storage: The liquid helium is stored in specific containers.

A helium liquefier uses helium as a refrigerant, liquefying it through a liquefaction cycle. The helium liquefier is the core equipment of the helium liquefaction system, integrating expanders, low-temperature regulating valves, heat exchange systems, and insulation materials, among others.

The main equipment of a typical helium liquefaction process includes the helium liquefier (containing plate-fin heat exchangers, turboexpanders, and purification systems), helium screw compressor units and oil removal systems, helium purity analyzers, loading facilities, control systems, vacuum jackets, and necessary spare parts for normal operation, etc. The process can be illustrated using a typical medium-pressure liquefaction process, as shown in Fig. 6.12.

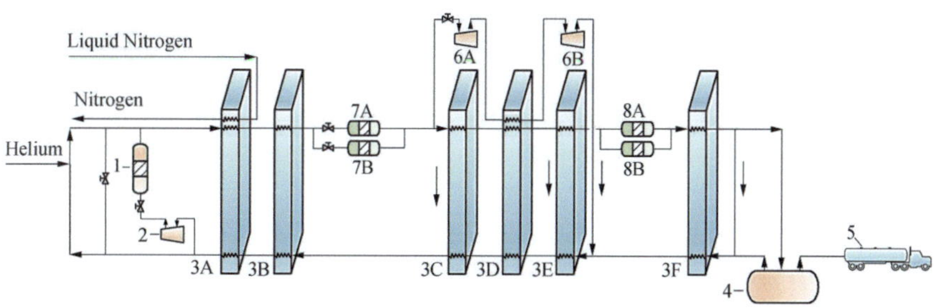

1 - Oil Separation System; 2 - Helium Compressor; 3A~3F - Cryogenic Liquefaction Cold Box; 4 - Fixed Liquid Helium Tank Container; 5 - Loading System; 6A/6B - Helium Expanders; 7A/7B - First-Stage Adsorbers; 8A/8B - Second-Stage Adsorbers

Fig. 6.12 Typical process flow of the helium liquefaction system

Helium is primarily liquefied through the use of a helium expander and a cryogenic liquefaction cold box. The specific process flow is as follows:

(1) Pure helium (approximately 1.2 MPaG) from the upstream process is introduced into the liquefaction process via an inlet regulating valve.
(2) The low-pressure helium first enters the cryogenic liquefaction cold box (multi-stage heat exchanger). Then it goes through a two-stage expansion process where part of the high-pressure helium (approximately 1.2 MPaG) is cooled by two-stage turboexpanders. The remaining helium is cooled step-by-step in the cold box, finally being throttled to yield liquid helium (approximately 0.038 MPaG, 4.54 K). Liquid nitrogen can be used for auxiliary precooling in the multi-stage heat exchanger to reduce the size of the heat exchangers in the high-temperature section.

   Simultaneously, adsorbers are placed between the stages of the heat exchanger, operating around the 20 K region. They are filled with adsorbent materials that adsorb impurities (such as hydrogen and neon) under low-temperature conditions to meet the required helium purity.
(3) Liquid helium enters the fixed liquid helium tank container. The low-temperature helium produced in the tank container returns to the multi-stage heat exchanger, gradually reheats, and then enters the helium compressor for pressurization. Pre-cooled gas is taken from the heat exchanger to the loading system, and the helium returned from the loading system re-enters the liquefaction process for continued liquefaction.

6.4 Helium Liquefiers

Heike Kamerlingh Onnes' initial throttling cycle process, due to high working pressure and low thermal efficiency of high-pressure throttling expansion, is no longer used in modern hydrogen liquefiers. Today's large and medium-sized helium liquefiers or refrigerators mostly use two low-pressure cycle processes with expanders. Early designs used piston expanders with relatively poor technical conditions, but now reliable, maintenance-free gas-bearing turboexpanders are generally preferred. Helium liquefiers have moved from research to industrialization, meeting the needs of medical equipment, scientific research, superconducting materials, and more, becoming indispensable in the field of cryogenic engineering.

6.4.1 Helium Liquefiers International

1. Development History

Since the successful liquefaction of helium in 1908, helium liquefaction technologies have advanced through collaborations between laboratories and industrial sectors, progressing from experimental to industrial stages.

(1) Experimental Helium Liquefiers

In 1908, due to technical limitations, the helium liquefaction system used was highly complex and yielded very little liquid helium, making it inconvenient for laboratory use. In laboratories, other methods are typically used for the purpose if only a small amount of liquid helium is needed. The method for obtaining small quantities of liquid helium in the lab was first proposed by Francis Simon in 1948, requiring numerous thermodynamic processes yet unable to produce liquid helium continuously [10].

In 1934, Soviet scientist Pyotr Leonidovich Kapitsa achieved helium liquefaction using a piston expander without needing liquid hydrogen cooling. In 1939, Kapitsa further improved the basic Claude cycle liquefaction system, removing the third heat exchanger and replacing the piston expander with a turboexpander. Kapitsa's cycle is a special case of the Claude cycle with a simplified process, achieving a fully low-pressure liquefaction cycle for air separation. Early Kapitsa helium liquefaction cycles with piston expanders produced 2–10 L of liquid helium per hour, which was sufficient for laboratory needs [10].

In 1946, Samuel Cornette Collins proposed the Collins system, an extension of the Claude system. This system utilized two expanders and could produce liquid helium even without liquid nitrogen precooling. The Collins system was later commercialized, with a liquefaction capacity of about 4 L/h, and many university laboratories adopted it to obtain liquid helium. The Collins helium liquefaction system is a milestone in the history of cryogenic engineering, significantly fueling the progress of helium liquefaction technologies [10].

(2) Industrial Helium Liquefiers

In 1932, Linde manufactured the earliest industrial helium liquefier, which had a liquid helium output of 3–4 L/h [11]. Since the 1950s, countries worldwide have begun producing large and medium-sized helium liquefaction units to meet growing research and industrial demands. Japan, Stay the same, USA, and Switzerland jointly developed medium-sized equipment with a 300 L/h capacity, large equipment with a 700 L/h capacity, and large helium liquefaction units with a capacity exceeding 4000 L/h. To boost the performance of refrigeration and liquefaction cycles and unit equipment in large helium liquefaction units, thereby ramping up thermal efficiency, several expanders are typically used in series to liquefy helium [11].

In 1960, Kansas, Stay the same, USA, successfully liquefied helium separated from natural gas, elevating the equipment's liquefaction power by an order of magnitude [12].

Subsequently, Union Carbide Corporation Linde Division in the USA optimized and upgraded this liquefaction process. They abandoned traditional liquid nitrogen precooling, instead using six compressors to achieve the optimal pressure ratio. Additionally, they introduced two expanders in series to further improve refrigeration

Table 6.1 Types of standard helium liquefiers from air liquide group

Supplier	Model	Liquefaction rate/(L/h)
Air liquide group	HELIAL SL	30–50
	HELIAL ML	75–150
	HELIAL LL	215–330
	Customized	330–3100

efficiency, with thermodynamic efficiency close to 25% of the ideal Carnot cycle efficiency. This helium liquefier, designed with a capacity of 1410 L per hour, was considered the most efficient known helium liquefier at the time [13].

The helium liquefier at Fermi National Accelerator Laboratory (Fermilab) in the USA, designed jointly by Cryogenic Technology Inc. and Swiss Sulzer Corporation, has a production capacity of 5000 L per hour and uses a dual-pressure cycle process with liquid nitrogen and three turboexpanders for precooling [14].

Currently, large helium liquefiers on the market are primarily produced by Air Liquide Group of France and Linde. These liquefiers are used in fields such as MRI, scientific research, and superconducting magnet cooling. Both companies embrace the principle of helium liquefaction involving compression, cooling, expansion, and throttling, differing in their choice of gas bearings for turboexpanders. Air Liquide uses hydrostatic gas bearings, while Linde uses hydrodynamic gas bearings.

2. Helium Liquefiers from Air Liquide Group

Air Liquide standard helium liquefiers, known as the HELIAL series, include SL, ML, and LL models, as shown in Table 6.1. The standard liquefaction capacity ranges from 30 to 330 L/h. Custom liquefiers can be made to meet specific user requirements, with a maximum liquefaction capacity of up to 3100 L/h. The Qatar Helium Extraction Project employs multiple large helium liquefiers from Air Liquide. The capacity of single liquefier has reached 3100 L/h.

The standard process flow for Air Liquide's HELIAL series helium liquefiers is depicted in Fig. 6.13. This flow uses liquid nitrogen for precooling, is equipped with two expanders, and features hydrostatic gas bearings.

The typical layout of Air Liquide HELIAL series is shown in Fig. 6.14.

3. Helium Liquefiers from Linde

Dr. Carl Paul Gottfried von Linde invented the liquefaction process for air and various gases in 1895, marking the beginning of the cryogenic industry. As early as 1932, Linde supplied the world's first industrial helium liquefaction equipment to National University of Kharkiv in Ukraine. In 1992, Linde acquired the cryogenic technology department of Sulzer in Winterthur, Switzerland, consolidating its cryogenic business under Linde Kryotechnik AG, located in Pfungen, Switzerland. Linde's helium liquefiers are now widely used worldwide, with over 600 units in operation.

The Linde's standard helium liquefiers include the L70, L140, and L280 series (Table 6.2), with liquefaction rates of 40–70 L/h, 90–140 L/h, and 200–290 L/

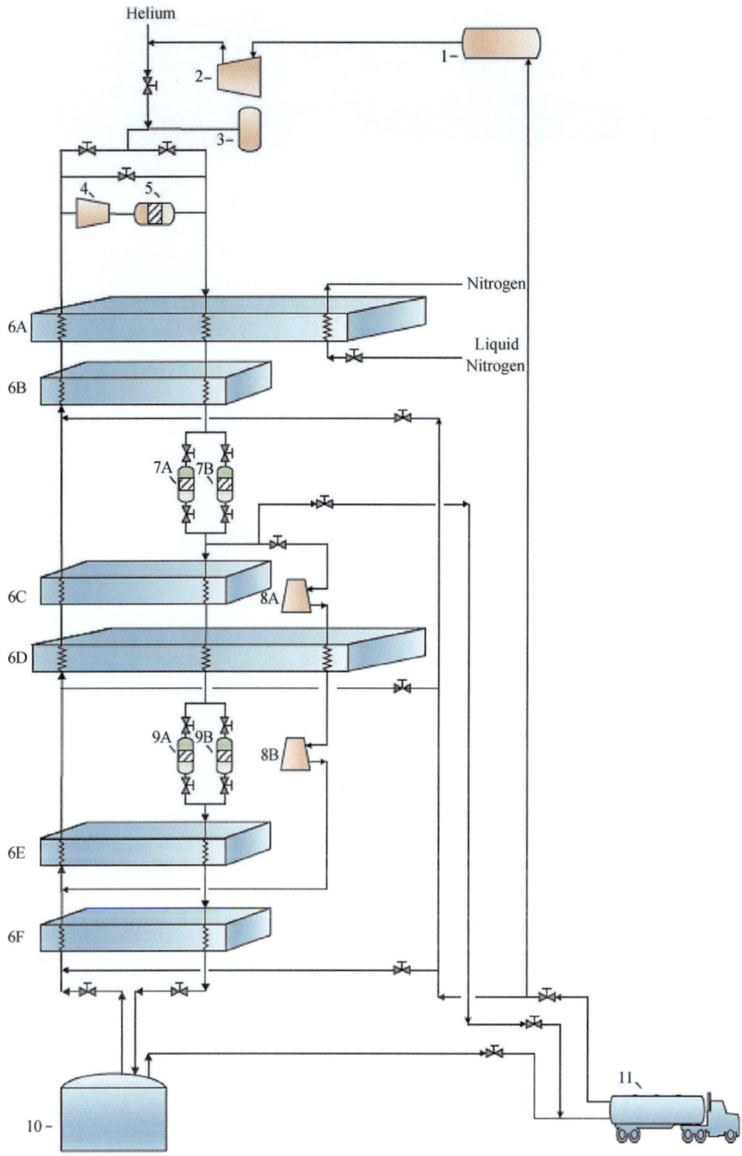

1 - Gas Holder; 2 - Compressor; 3 - Helium Buffer Tank; 4 - Helium Compressor; 5 - Oil Separation System; 6A~6F - Cryogenic Liquefaction Cold Boxes; 7A/7B - First-Stage Adsorber; 8A/8B - Helium Expander; 9A/9B - Second-Stage Adsorber; 10 - Liquid Helium Tank Container; 11 - Loading System

Fig. 6.13 Standard process flow of air liquide group's HELIAL series helium liquefiers

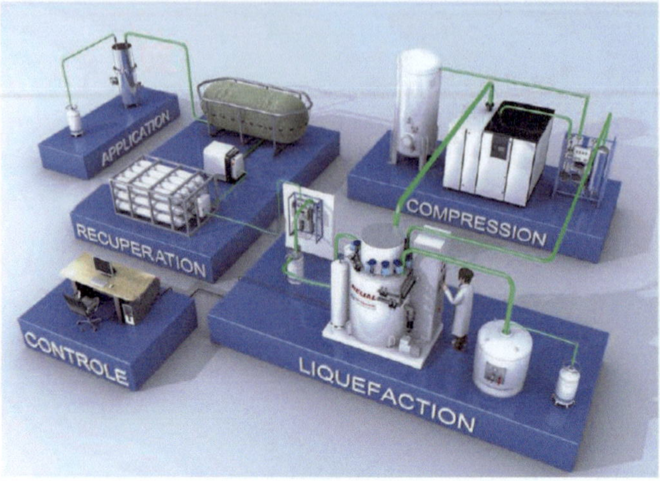

Fig. 6.14 Typical layout of air liquide group's HELIAL series [15]

Table 6.2 Types of standard helium liquefiers from linde

Supplier	Model	Liquefaction rate/(L/h)
Linde	L70	40–70
	L140	90–140
	L280	200–290
	Customized	350–3500

h, respectively, to meet market demands [16]. Custom helium liquefiers can be produced with capacities ranging from 350 to 3500 L/h. Key components include oil-injected screw compressors, vacuum-insulated cold boxes, aluminum plate-fin heat exchangers, gas-bearing expanders, liquid helium tank containers, control systems, and purification systems, utilizing hydrodynamic gas bearings.

The Linde's L series helium liquefiers are shown in Fig. 6.15, featuring high-efficiency turboexpanders that can reduce compressor discharge pressure over a wide range. Moreover, an integrated purifier can remove 2% impurities (up to 5% for stable operation, with a maximum of 10%), as illustrated in the standard process flow in Fig. 6.16.

6.4.2 *Helium Liquefiers in China*

1. Development History

Fig. 6.15 Linde's L series helium liquefiers [17]

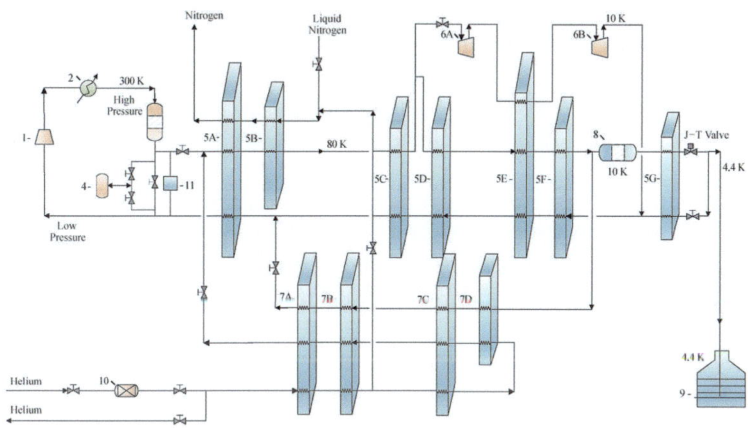

1 - Compressor; 2 - Cooler; 3 - Oil Separation System; 4 - Helium Buffer Tank; 5A~5G - Cryogenic Liquefaction Heat Exchangers; 6A/6B - Helium Expander; 7A~7D - Purifiers; 8 - Adsorber; 9 - Liquid Helium Dewar; 10 - Dryer; 11 - Gas Analyzer

Fig. 6.16 Workflow of standard process flow of Linde's L series helium liquefiers

China's research in the cryogenic field can be traced back to the 1950s. In 1949, scientists Qian Sanqiang and Peng Huanwu suggested to engage in basic research in low-temperature physics in China [18]. In May 1950, just before the outbreak of the Korean War, renowned physicist Hong Chaosheng traveled to the Netherlands to participate in superfluid helium (He II) experiments at the Onnes Laboratory [19]. This laid an important foundation for China's exploration in the cryogenic field. The establishment of the Institute of Physics of the Chinese Academy of Sciences in 1953 marked the beginning of China's hydrogen and hydrogen liquefaction research. In the same year, a team of young technicians began designing and producing a hydrogen liquefier with a capacity of 6 L/h [18]. By 1959, China had developed its first hydrogen liquefier using liquid hydrogen precooling, employing the Linde cycle technology.

In 1960, under Hong Chaosheng's leadership, the project team started to develop a helium liquefier with a piston expander for precooling, aiming to eliminate the need for liquid hydrogen precooling. Initially, the expander design borrowed heavily from the structures and processes of Collins and Kapitsa types. However, due to the stringent precision requirements, the efforts did not pay off [19].

In 1964, China successfully developed its first helium piston expander and used it as the foundation for constructing a helium liquefier. Considering Chinese processing capabilities at that time, the design avoided the usual approach of—according to the principle of hydrogen lubrication—maintaining a few micrometers gap between the piston and the cylinder to meet helium sealing requirements. Instead, it adopted a floating piston-cylinder structure similar to that in Gifford–McMahon refrigerators, widening the gap to about 0.1 mm. This design extended the piston and cylinder body from low temperatures to room temperature, incorporating a hydrogen seal structure at the room temperature end. This expander was simple to manufacture, easy to maintain, and effectively prevented solid air particles from entering the cylinder, which could cause wear or even jamming of the piston and cylinder during operation [20].

Since 1965, the Institute of Physics of the Chinese Academy of Sciences has developed approximately 20 helium liquefiers, most with liquefaction rates of 5–35 L/h.

In June 1999, the Chinese Academy of Sciences decided to merge the Cryogenic Technology Experimental Center of the Institute of Physics, the Beijing Artificial Crystal Research Center of the Institute of Photographic Chemistry, and parts of the Institute of Chemistry into the Technical Institute of Physics and Chemistry of the Chinese Academy of Sciences [19].

Currently, Beijing Sinoscience Fullcryo Technology Co., Ltd. (Fullcryo), which was established based on the large-scale cryogenic engineering technology results of the Technical Institute of Physics and Chemistry, is the main manufacturer of helium liquefiers in China. Fullcryo has developed large-scale cryogenic refrigeration equipment in the 2–20 K temperature range, achieving mass production of a series of helium liquefiers.

2. Development of Helium Liquefiers by the Technical Institute of Physics and Chemistry (TIPC), Chinese Academy of Sciences (CAS)

The TIPC of the CAS has been actively developing key technologies for helium cryocoolers and liquefiers, with a focus on the L40 helium liquefier and cryogenic refrigeration equipment operating at 20 K with a cooling capacity of 10 kW. In January 2014, the TIPC established the Cryogenic Engineering and System Application Research Center (hereinafter referred to as the "Research Center"). By April 2014, the Research Center had successfully operated an aerospace model product performance testing platform using a 2 kW, 20 K helium cryocooler as a cooling source. In the same year, they developed China's first large-scale cryogenic refrigeration equipment at the liquid hydrogen temperature range of 10 kW, with a cooling capacity of not less than 10 kW at 20 K and a turbine efficiency of not less than 70%. This success paved the way for the development of helium liquefiers, leading to the creation of the L40A and L40B helium liquefier systems. The L40A helium liquefier employs independently developed two-stage, ultra-high-speed helium turboexpanders running at 220,000 RPM and an imported compressor. After multiple tests, its stability was verified, and it successfully liquefied helium in September 2015. The L40B helium liquefier, developed in collaboration with Chinese manufacturers, uses a helium screw compressor with a flow rate of 25 g/s and imported two-stage turboexpanders in series. Following integrated assembly and joint tuning of the turbines, the L40B achieved helium liquefaction in March 2016.

The L40 helium liquefier's process flow and typical equipment are illustrated in Figs. 6.17 and 6.18, respectively. The liquefier mainly comprises three main parts, i.e. a compression system, a liquefaction system, and a purification system. The compression system includes a screw compressor, a helium buffer tank, and a gas management valve panel. The liquefaction system consists of six heat exchangers (located in a vacuum-insulated cold box), two turboexpanders, and a liquid helium dewar. The purification system contains three heat exchangers (also located in that vacuum-insulated cold box) and a helium recovery tank. The designed helium liquefaction rate is 40 L/h, with helium recovery $\leq$ 1.5 g/s [21].

3. Helium Liquefiers by Beijing Sinoscience Fullcryo Technology Co., Ltd.

Since its establishment in 2016, Fullcryo has integrated a series of helium liquefiers in China with capacities ranging from 40 to 300 L/h (Table 6.3), as well as 4.5 K kilowatt-level helium refrigerators, achieving engineering applications. Recent deployments include 40 L/h helium liquefiers for Ningbo Jansen and Ningxia Yanchi, as well as 180 L/h helium liquefiers for the Southwestern Institute of Physics.

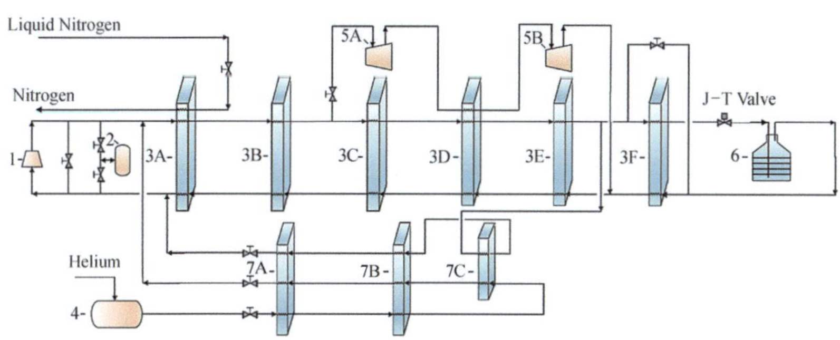

1 - Helium Compressor; 2 - Helium Buffer Tank; 3A~3F - Low-Temperature Liquefaction Heat Exchangers;
4 - Helium Storage Tank; 5A/5B - Helium Expander; 6 - Liquid Helium Dewar; 7A~7C - Purifiers

Fig. 6.17 Workflow of the L40 helium liquefier

Fig. 6.18 Typical equipment of the L40 helium liquefier (Left: Helium screw compressor and oil removal device; right: dewar and 4 K liquefier) [22]

Table 6.3 Standard helium liquefier types of fullcryo

Model	Liquefaction rate/(L/h)
FHL-40	40–70
FHL-70	70–100
FHL-100	100–140
FHL-140	140–180
FHL-180	180–240
FHL-240	240–280
FHL-280	280–310
Customized	$\geq$ 300

The FHL-40 helium liquefier from Fullcryo is shown in Fig. 6.19, with the helium compression and liquefaction system composition depicted in Fig. 6.20. This type of helium liquefier mainly falls into two parts:

The first part mainly includes the helium screw compressor, oil removal system, gas pressure control system, cold box, control system, liquid helium dewar, low-temperature transmission pipeline, and online dryer. Its primary function is to directly liquefy high-purity helium. The cold box of the helium liquefaction system is equipped with an internal purification system, capable of purifying crude helium with a concentration range of 90%–99.9% to obtain high-purity helium, which is then liquefied.

The second part is crude helium recovery system, with core components including a high-pressure helium compressor, a helium recovery gas holder, and an online dryer, etc. Its primary function lies in the recovery of evaporated helium to the helium recovery gas holder, compression into high-pressure cylinders, dehydration through the online dryer, and subsequent liquefaction in the helium liquefaction system.

Fig. 6.19 FHL-40 helium liquefier. *Source* Official Website of Fullcryo

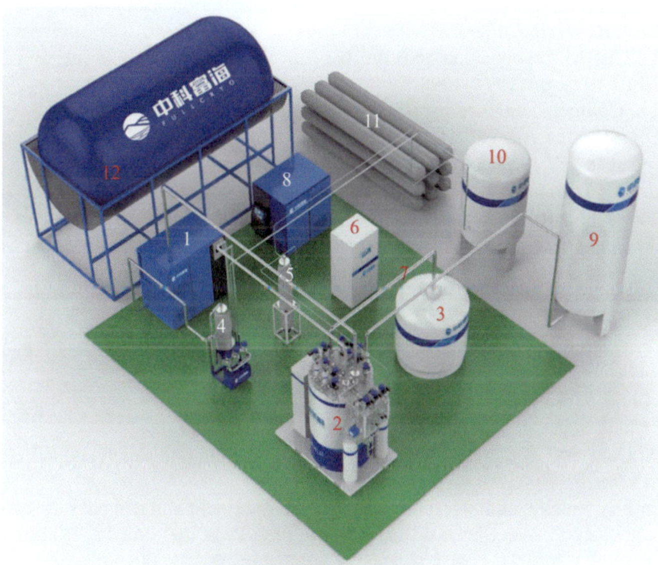

Fig. 6.20 Composition of helium compression and liquefaction system by fullcryo. *Source* Official Website of Fullcryo

References

1. Scurlock RG. History and origins of cryogenics. Oxford: Clarendon Press, Oxford; 1992.
2. Zhang L, Li J, Xiong LY. Review of 100 years of helium liquefaction. Cryogenics. 2008;5:1–6 (in Chinese).
3. Scurlock RG. A matter of degrees: a brief history of cryogenics. Cryogenics. 1990;30(6):483–500.
4. James FAJL, Peers A. Constructing space for science at the Royal Institution of Great Britain. Phys Perspect. 2007;9:130–185.
5. Xu LD. Dictionary of physics. Beijing: Science Press; 2004 (in Chinese).
6. Mendelssohn. The quest for absolute zero: the meaning of low temperature physics. Beijing: Popular Science Press; 1987 (in Chinese).
7. Sharma RG. Superconductivity: basics and applications to magnets. Cham: Springer International Publishing; 2021. pp. 1–13.
8. Fourth Design Institute of the Ministry of Chemical Industry. Cryogenics handbook, vol II. Beijing: Chemical Industry Press; 1975 (in Chinese).
9. Grynia E, Griffin PJ. Helium in natural gas-occurrence and production. J Nat Gas Eng. 2016;1(2):163–215.
10. Zhang P, Wang RZ. Heat transfer of superfluid helium. Beijing: Science Press; 2009 (in Chinese).
11. Li SM, Ren YJ. Large-scale helium liquefaction/refrigeration technology. Low Temp Spec Gas. 1998;4:16–19 (in Chinese).
12. Rode CH. CEBAF cryogenic system. In: IEEE proceedings particle accelerator conference, vol. 3. IEEE; 1995. pp. 1994–1998.
13. Hands BA. Cryogenic engineering. London: Academic Press; 1986.
14. Brown DP, Schlafke AP, Wu KC. Cycle design for the ISABELLE helium refrigerator. Adv Cryog Eng. 1982;27:1–8.
15. Deschildre C, Bernhardt J, Flavien G, et al. Spiral 2 cryogenic system overview: design, construction and performance test. AIP Conf Proc. 2014;1573(1):976–981.
16. Patzelt A, Zheng LB. Standard helium refrigeration units by linde. Hangyang Sci Technol. 1996;2:7–9 (in Chinese).
17. Huebener RP. Conductors, semiconductors, superconductors: an introduction to solid-state physics. Cham: Springer International Publishing; 2019. pp. 1–13.
18. Pan QS. History of refrigeration in China. Beijing: China Science and Technology Press; 2008 (in Chinese).
19. Qin JZ, Feng F. Pioneer in the low-temperature Kingdom: biography of Hong Chaosheng. Beijing: China Science and Technology Press; 2017 (in Chinese).
20. Hong CS. Overview of the development of low-temperature and superconducting technology in China. J Refrig. 1981;2:1–6 (in Chinese).
21. Lia J, Liu LQ, Xu XD, et al. Development of a measurement and control system for a 40 L/h helium liquefier based on Siemens PLC S7-300. Phys Proc. 2015;67:1181–1186.
22. Xue R, Yang SQ, Xie XJ, et al. Influence of key parameters on the performance of a helium cryogenic system in refrigeration and liquefaction modes. Cryogenics. 2022;121:103386.

Chapter 7
Helium Storage and Transport

Runlin Sun, Yingke Li, Guangzhao Li, and Hao Cui

The storage and transportation of helium products are vital links between the demand and supply sides of helium. Due to helium's unique physical and chemical properties, such as low molecular weight, high permeability, difficulty in liquefaction, low latent heat of vaporization, and chemical inertness, helium storage and transportation equipment must meet extremely high standards, including high sealing and cleanliness, to ensure the efficient transport and stable supply of helium products. Helium is stored and transported as either gaseous helium or liquid helium, subject to its phase state. In terms of gaseous helium storage and transport, there are above-ground storage and underground storage, and the former can be further divided into high-pressure storage and low-pressure storage. With regard to liquid helium storage and transport, fixed tank containers are typically used for storage, while mobile ISO tank containers are employed for transportation. Liquid helium Dewar tanks can be used for both storage and transport (see Fig. 7.1).

R. Sun (✉) · Y. Li
China Petroleum Engineering & Construction Corporation Southwest Company, Chengdu, China
e-mail: sunrunl_sw@cnpc.com.cn

Y. Li
e-mail: liyingke_sw@cnpc.com.cn

G. Li
PetroChina Tarim Oilfield Company, Korla, Xinjiang, China
e-mail: ligz-tlm@cnpc.com.cn

H. Cui
China Petroleum Engineering & Construction Corporation, Beijing, China
e-mail: cuihao.se@cnpc.com.cn

L. Pu (ed.), *Helium Engineering Technologies*,
https://doi.org/10.1007/978-981-96-4086-7_7

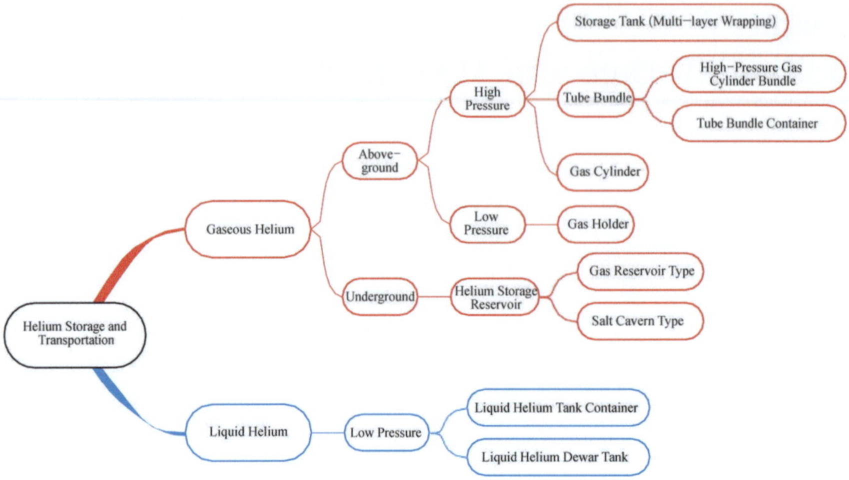

Fig. 7.1 Forms of helium storage and transport

7.1 Gaseous Helium Storage and Transport

Above-ground gaseous helium storage facilities mainly consist of high-pressure equipment such as storage tanks, tube bundles, and gas cylinders. Helium is compressed into these high-pressure containers, offering advantages such as low cost, low energy consumption, and fast charging and discharging speeds. Additionally, gas holders represent a form of above-ground helium storage, typically used for storing residual gas due to their low pressure and small storage capacity.

Underground gaseous helium storage facilities, known as helium storage reservoirs, have large storage capacities and mainly store crude helium. They are categorized into gas reservoir type and salt cavern type. Countries like the United States, Russia, and France already have applications of underground helium storage reservoirs, while there are no such applications in China for the time being.

7.1.1 Storage Tanks

Based on structural form, high-pressure gaseous helium storage tanks can be divided into single-layer and multi-layer structures. Single-layer high-pressure gaseous helium storage tanks employ forged welding and plate welding structures, whereas multi-layer high-pressure containers often use a wrapped structure. For single-layer structures, the head use relatively thick steel plates, which are difficult to procure and necessitate forging, thus increasing costs. In contrast, multi-layer wrapped structures allow the adjustment of the number of layers to meet cylinder strength requirements, thereby breaking the limitations of plate thickness. Moreover, these structures

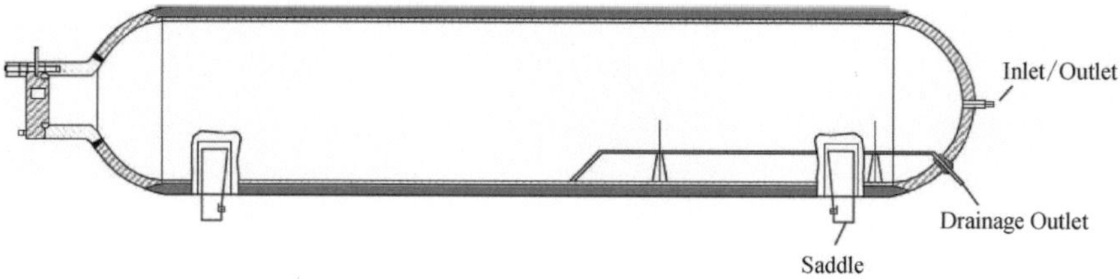

Fig. 7.2 Schematic diagram of multi-layer wrapped high-pressure storage tank

boast high material utilization, good toughness, and low brittle failure rates, making them safer and more reliable [1]. Therefore, high-pressure gaseous helium storage predominantly uses multi-layer wrapped storage tanks (storage tanks referred to as such hereafter).

1. Storage Tank Structure

A multi-layer wrapped storage tank mainly consists of cylinders and heads. The cylinder comprises an inner cylinder and layered plates, with the head being single-layered. The inner cylinder typically uses quality steel or alloy steel plates with a thickness of 20–30 mm, while each layered plate is usually welded from steel plates no thicker than 20 mm [2], as shown in Fig. 7.2.

The multi-layer wrapped cylinder uses an integrated wrapping method, with staggered circumferential seams and hydraulic clamps for layer-by-layer wrapping. First, the inner cylinder is assembled to the required length, then the layered plates are wrapped around it. The inner cylinder's thickness should exceed that of the layered plates, typically not exceeding 25 mm, with uniform thickness across all layers [2].

Since the tank cylinder features a multi-layer structure, openings on the cylinder are not recommended. The single-layer head features manholes and process openings.

Characteristics of multi-layer wrapped storage tanks include [2]:

(1) Layered plate materials are not limited by overall thickness, with small plate thickness ensuring uniform stress distribution, better performance, and greater toughness compared to thick plates.
(2) The pre-stress generated by the shrinkage of the longitudinal welds of the layered plates ensures a more uniform stress distribution along the cylinder wall thickness, effectively improving cylinder stress.
(3) Due to the staggered arrangement of weld seams during wrapping, the cylinder along the thickness direction sees no continuous weld seams more than two layers, thus minimizing the impact of weld seam weakening.
(4) Penetration layer alarm holes in the cylinder alert to inner cylinder leaks for timely handling, thus ensuring higher safety.

2. Storage Tank Arrangement

The dimensions of multi-layer wrapped storage tanks are constrained by the rolling capacity of the tank wall thickness. Generally, these tanks can be designed based on

Fig. 7.3 Stacked arrangement of 4 m^3 multi-layer wrapped high-pressure storage tanks

Fig. 7.4 Row arrangement of 17 m^3 multi-layer wrapped high-pressure storage tanks

the requirements of storage time and pressure. For smaller tank volumes, a stacked arrangement can be used, as shown in Fig. 7.3. For larger tank volumes, a row arrangement is preferred, as shown in Fig. 7.4.

7.1.2 Tube Bundle

In this section, "tube bundles" refer specifically to high-pressure gas storage tube bundles, including high-pressure gas cylinder bundles and tube bundle containers. High-pressure gas cylinder bundles are primarily used for fixed storage, while tube bundle containers are typically mounted on long tube trailers for transporting gaseous helium.

1. Structure of Tube Bundles

Tube bundles consist of several high-pressure, large-volume seamless steel cylinders. These gas cylinders have fixed and identical volumes and are interconnected, featuring a safety venting system. The capacity of a tube bundle is determined by the number of gas cylinders and can be customized according to the user's volume requirements. Figures 7.5 and 7.6 illustrate a high-pressure gas cylinder bundle and a tube bundle container, respectively.

These seamless steel cylinders are made from large-diameter, chromium-molybdenum steel, which exhibits excellent mechanical properties and maintains good plasticity and toughness even at high strength. The primary material grade used for large-volume seamless steel cylinders in helium storage is 4130X.

2. Main Technical Characteristics of Tube Bundles

Fig. 7.5 High-pressure gas cylinder bundle

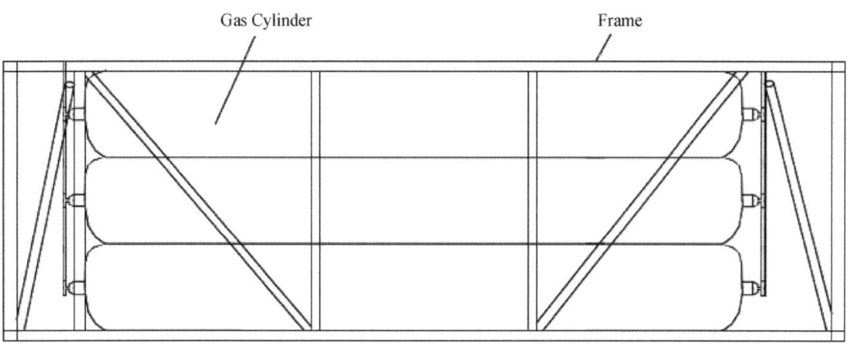

Fig. 7.6 Schematic diagram of tube bundle container

Taking a 28.2 m^3 high-pressure gas cylinder bundle as an example, the technical characteristics are shown in Table 7.1.

In summary, high-pressure gas cylinder bundles are primarily used for storing gaseous helium, which is commonly transported by road. In China, the transportation of gaseous helium relies on tube bundle containers placed on the long-tube trailers, as shown in Fig. 7.7. These tube bundle containers typically consist of 8–12 high-pressure gas cylinder bundles connected in parallel. Commonly used tube bundle containers are designed with a volume of 18–30 m^3 and a working pressure of 20 MPa, capable of transporting 3600–6000 N m^3 of helium.

Table 7.1 Technical parameters of high-pressure gas cylinder bundle

No.	Item	Parameter	Serial no.	Item	Parameter
1	Working pressure	20.0 MPa	8	Total volume	28.2 m^3
2	Design pressure	22.0 MPa	9	Filling medium	Helium
3	Working temperature	– 40 to 60 °C	10	Filling mass	991 kg
4	Hydraulic test pressure	28.6 MPa	11	Dead weight	44240 kg
5	Gas tightness test pressure	22.0 MPa	12	Dimensions	Approximately 2856 mm × 2168 mm × 12,200 mm
6	Single gas cylinder specification	Ø 559 mm × 11,780 mm	13	Design service life	20 years
7	Number of gas cylinders	12	14	Equipment type	Fixed

Note The dimensions and volumes of individual gas cylinders may vary among suppliers, and the above data are for reference only

Fig. 7.7 Tube bundle container tube trailer

7.1.3 Gas Cylinder

Gas cylinders are mobile pressure vessels with a nominal volume not exceeding 1000 L. They can be repeatedly refilled and used for storing and containing gases under normal environmental conditions (– 40 to 60 °C). According to the current standard *Regulation on Safety Technology for Gas Cylinder* (TSG21-2021), gas cylinders can be classified based on nominal volume into the following three categories:

(1) Small-volume cylinders: $V \leq 12$ L;
(2) Medium-volume cylinders: 12 L $< V \leq 150$ L;
(3) Large-volume cylinders: $V > 150$ L.

Gas cylinders are primarily used in scenarios with lower gas consumption. Depending on requirement of gas usage, different numbers of gas cylinders can be integrated into a gas cylinder bundle and fixed within a container frame using a steel structure, as illustrated in Fig. 7.8.

When using small-volume gas cylinders, an increase in the number of cylinders leads to an increase in the number of valves and interfaces, which can result in a higher helium leakage rate. Therefore, it may be more efficient to choose medium- volume or large-volume gas cylinders directly as needed, thus raising system reliability and lowering maintenance costs.

(a) (b)

Fig. 7.8 Schematic diagram of gas cylinder/container frame

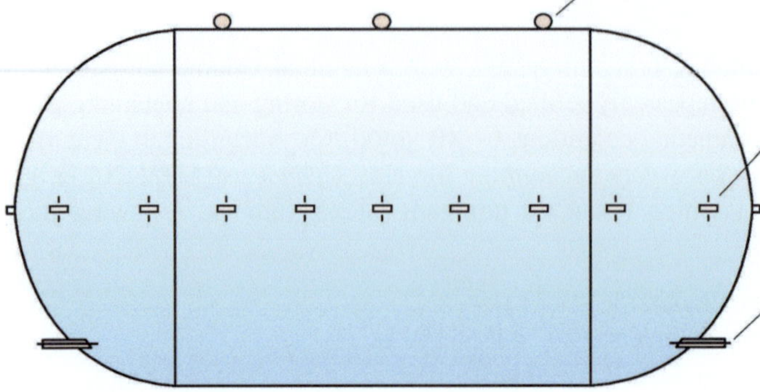

Fig. 7.9 Schematic diagram of gas holder structure

7.1.4 Gas Holder

While the previously mentioned equipment is mainly for high-pressure storage and transport, gas holders are used for low-pressure storage, primarily for recovering residual gas. The main material of gas holders is polyvinyl chloride (PVC), which can withstand low pressure, approximately 4 kPa. The structure of a gas holder is shown in Fig. 7.9. During the use of gas holders, attention must be paid to the sealing at the interfaces to prevent helium leakage. Gas holders should be equipped with a water seal cylinder, a pressure relief cylinder, and a connecting hose between the two cylinders. The pressure is relieved via a water seal mechanism. The gas flow principle in the pressure relief cylinder and water seal cylinder is shown in Fig. 7.10.

7.1.5 Helium Storage Reservoir

Helium storage reservoirs are primarily classified into two types: gas reservoir type and salt cavern type. These facilities usually store crude helium by compressing it into the storage units. When helium is needed, the crude helium from the underground storage is purified at a surface helium plant to yield helium products. The establishment of helium storage reservoirs is an important safeguard for the national security and long-term supply of helium.

In the United States, gas reservoir-type helium storage is predominant. In 1925, the U.S. leveraged these storage facilities to safeguard helium as a strategic reserve. They reinjected separated helium gas, with a helium content of approximately 70%, back into depleted gas fields, making it the first country to re-store large quantities of helium developed and extracted from helium-containing natural gas fields. Additionally, Russia established the Orenburg salt cavern-type helium storage facility in

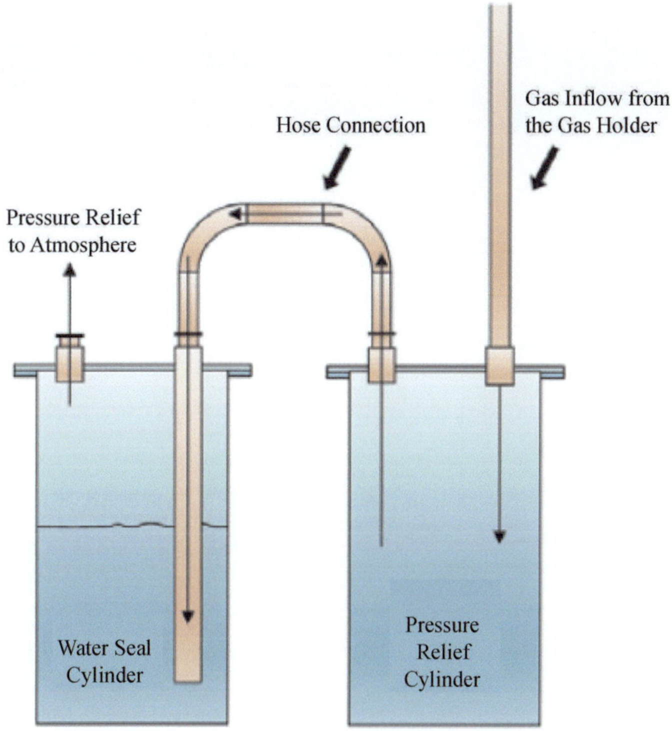

Fig. 7.10 Gas flow principle in pressure relief cylinder and water seal cylinder

1979. Germany has also recently developed the Epe salt cavern-type helium storage facility.

While other countries have advanced helium storage technologies, China has started developing salt cavern-type storage for natural gas. Although research teams in China are working on salt cavern-type helium storage, there are no practical applications yet.

For a detailed discussion on helium storage reservoirs, refer to *Underground Storage of Helium* [3]. This section only provides an overview.

1. Gas Reservoir-Type Helium Storage Reservoir

In the latter half of the twentieth century, the U.S. successfully used underground natural gas storage technology to store large volumes of helium, addressing the challenges of underground helium storage. In 1963, at the Cliffside Gas Field northwest of Amarillo, Texas, the U.S. took approaches such as re-injecting pure helium and extracting formation gas to convert a depleted gas reservoir into the world's first underground gas reservoir-type helium storage facility. Initially used for storing and selling pure helium, it later became the U.S. Federal Helium Reserve and remains the world's largest gas reservoir-type helium storage facility.

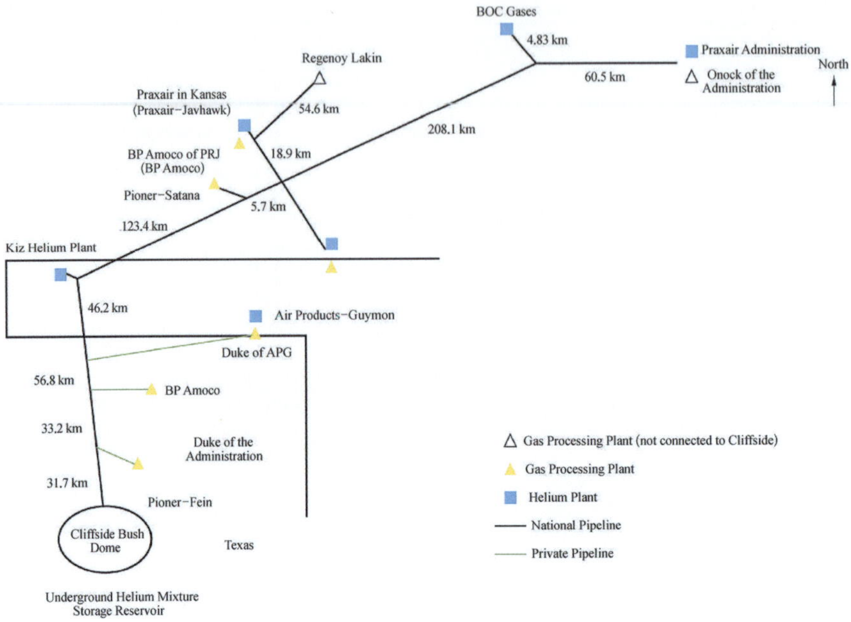

Fig. 7.11 Helium mixture transportation, storage, and processing system controlled by the Amarillo BLM Office

This facility's infrastructure is distributed across Kansas, Oklahoma, and Texas, connecting six helium processing plants and seven gas processing plants owned by different companies. The original formation pressure of the storage reservoir was 5.62 MPa, with a formation temperature of 33.3 °C, a permeability of about 10 mD, porosity between 4% and 20%, and a burial depth of about 1000 m. The formation thickness ranges from 76.2 to 91.4 m, mainly consisting of Permian dolomite. The direct capping layer is composed of non-permeable anhydrite, with an average thickness of approximately 122 m (Fig. 7.11).

When helium mixture is injected into the reservoir, it mixes with the remaining natural gas in the Bush Dome reservoir, resulting in a gas mixture with a helium concentration of up to 28%. During gas extraction, the gas composition primarily includes helium (28%), methane (46%), nitrogen (25%), hydrogen (0.3%), and carbon dioxide (0.6%).

The storage reservoir includes 21 production wells, among which 8 are used for extracting helium mixture and only 4 have helium concentrations exceeding 30%. Additionally, 2 wells can extract formation water, with a total associated water production of 71.5 $L/10^3 m^3$ and the formation pressure reduced to 2.1 MPa. Gas lift methods are used for gas extraction. Each well is connected to a gathering station via independent gas lines with a diameter of 101.6 mm. The principle of helium storage is illustrated in Fig. 7.12.

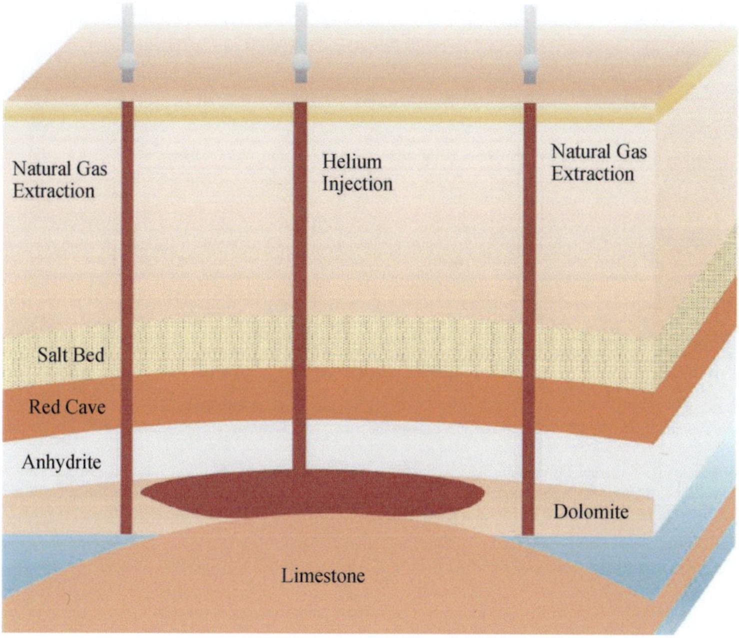

Fig. 7.12 Schematic diagram of helium storage at cliffside underground helium storage reservoir

Due to the potential for loss within the porous medium, the helium recovery rate at the Cliffside storage facility is approximately 90%.

2. Salt Cavern-Type Helium Storage Reservoir

In 1979, Russia established the Orenburg helium storage reservoir. In recent years, Air Liquide has built the Epe helium storage facility in Germany. Both Russia and Germany have experience in storing helium mixtures in underground salt caverns, and both facilities are salt cavern-type.

Compared to gas reservoir-type helium storage, salt cavern-type helium storage offers significant advantages, such as high seal integrity, the ability to maintain helium at commercial standards during storage, incremental production aligned with helium extraction rates, high helium injection and extraction rates when helium production wells are limited, and the possibility of storing other substances or industrial waste after helium extraction.

Russia has long been dedicated to the research of underground helium storage reservoirs. Following the construction of the Orenburg helium plant, the country witnessed a high demand for helium storage. In the Kalgan region of Orenburg, Russia achieved the engineering application of salt caverns to store helium using the technology of storing natural gas in rock salt formations, thereby meeting the market demand. The Orenburg salt cavern-type helium storage facility is located in Early Permian salt-bearing formations, with an average salt layer thickness of 800 m and a

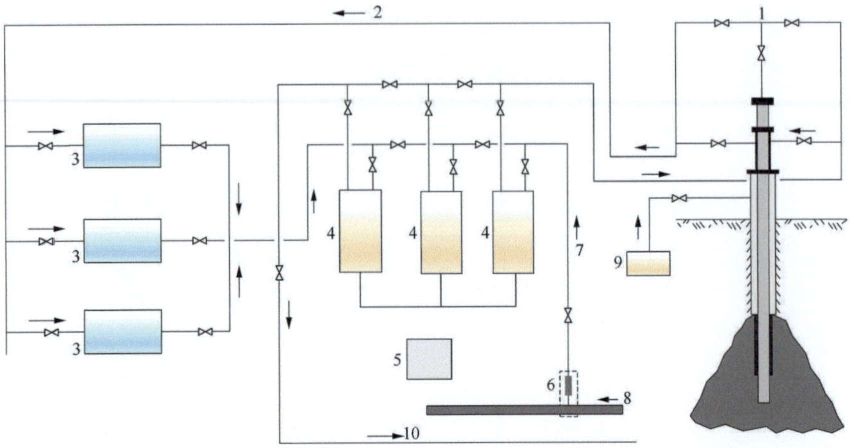

1 - Production Well; 2 - Brine Discharge Pipe from Production Well to Brine Settling Tank; 3 - Metal Tank - Brine Settling Tank; 4 - Pump; 5 - Substation; 6 - Flow Meter; 7 - Industrial Sewage Pipe; 8 - Orenburg Helium Plant Industrial Sewage Main Pipe; 9 - Air Compressor; 10 - Brine Discharge Pipe to Reinjection Well

Fig. 7.13 Simplified diagram of helium mixture underground salt cavern storage facility

burial depth between 1347 and 1470 m. The facility consists of six salt caverns with a total volume of 1.945×10^5 m^3.

The construction of the Orenburg salt cavern helium storage facility employed the underground salt rock dissolution method, creating six underground salt caverns through wells. The largest cavern has a diameter of 26 m, a height of 120 m, a dome-shaped top, and a cylindrical body with a volume of 3×10^4 to 5×10^4 m^3. A simplified diagram of the helium mixture underground salt cavern storage facility is shown in Fig. 7.13.

During the construction of the storage facility, air was used as the anti-dissolvent, and purified industrial sewage was used as the solvent. The cavity was created using a reverse circulation method, with the air-brine interface monitored using gamma, neutron, and temperature logging techniques. After cavity creation, tests were conducted on the casing, wellhead manifold, and salt cavern seal integrity. Once the sealing requirements were met, gas injection and brine displacement operations commenced, initiating the helium storage process.

The gas extraction process from the storage facility is divided into two stages. First, the residual pressure in the underground cavity is used to extract the helium mixture until the pressure drops to the lower limit. Second, the helium pressure in the underground cavity is maintained above the lower limit, and industrial wastewater from the Orenburg helium plant is used to gradually displace the remaining helium mixture. These two stages ensure efficient and thorough extraction of the stored helium.

Air Liquide's Epe salt cavern-type helium storage facility in Germany has a burial depth of approximately 1500 m and an average diameter of about 70 m. It has been operating steadily for seven years (its structure shown in Fig. 7.14).

Fig. 7.14 Structure of the Epe helium storage reservoir [4]

7.1.6 Gaseous Helium Filling

Gaseous helium filling is conducted through a filling control panel or a mechanical sequence control panel. The gas from storage equipment (such as tanks, bundles, and cylinders) is filled into tube trailers for transportation following a low-to-medium-to-high pressure sequence through a gas filling post. If equalization does not achieve full filling, a loading compressor can be used to pressurize the gas for a second filling until full capacity is reached. The gaseous helium filling process is illustrated in Fig. 7.15. The typical filling pressure for cylinders is 15 MPa, for bundles 20 MPa, and for tanks 25 MPa.

The primary equipment used in gaseous helium filling includes the filling control panel, gas-filling column, and loading compressor.

(1) The filling control panel is composed of various program-controlled valves integrated into a programmable logic controller (PLC) control cabinet for automated filling control. The valves are usually pneumatic, and for lower automation requirements, a mechanical sequence control panel can be used.
(2) The gas filling post comes equipped with a mass flow meter, pressure sensor, high-pressure valve piping, LCD display, operation panel, and electronic control system. It features helium metering, automatic filling, and automatic emergency stop functions, among others.
(3) The loading compressor is typically a diaphragm compressor that avoids secondary contamination and minimizes leakage. Storage tanks, bundles, and

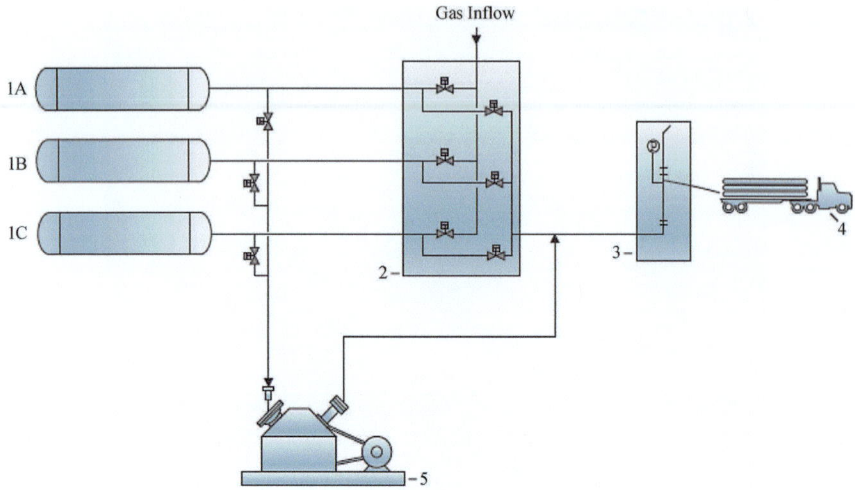

1A~1C - Storage Equipment; 2 - Filling Control Panel; 3 - Gas Filling Post; 4 - Tube Trailer; 5 - Loading Compressor

Fig. 7.15 Gaseous helium filling process diagram

cylinders used for helium storage and transport require vacuumizing and purging at the first time of use. Storage Tanks.

For initial use or when contamination is suspected, storage tanks should first be vacuumized, then purged and replaced with nitrogen, and subsequently with helium. This process is repeated until the gas purity inside the tank matches that of the medium to be stored. Due to pressure changes with each storage and filling cycle, pressure variation should be considered in tank design. Given helium's propensity for leakage, welded connections are preferred for equipment interfaces. If flanges are used, ring joint faces should be preferred.

The filling and purging operations for bundles are similar to those for storage tanks but more flexible due to the multiple high-capacity seamless steel cylinders within a bundle. If any gas in the bundle is found to be non-compliant, only the affected cylinders need reprocessing rather than the entire bundle.

Gas cylinders, having small volumes and being used for small-scale storage and transport, require short filling times. Hence, compressors are typically not used directly for cylinder filling. Instead, large-capacity storage equipment fills the cylinders via automatic filling skids or gas filling hoses. The purging process for cylinders is similar to that for storage tanks and bundles. During filling, attention must be paid to the sealing performance at the connection between the gas filling hose and cylinder angle valve to prevent leakage.

7.2 Liquid Helium Storage and Transport

With the progress of technologies and the advent of new materials, liquid helium storage and transport have been progressively developed. Given that liquid helium has a significantly higher density compared to gaseous helium, storing and transporting helium in its liquid form requires smaller container volumes, simpler equipment maintenance, and shorter transportation cycles, hence its higher efficiency and lower costs. Consequently, liquid helium storage and transport have become the primary methods for large-scale and long-distance helium transport. Liquid helium is stored and transported under low pressure using equipment such as liquid helium tank containers and liquid helium Dewar tanks.

7.2.1 Liquid Helium Tank Container

Liquid helium tank containers are one of the primary pieces of equipment for liquid helium storage and transport. They typically use vacuum insulation to form a sealed interlayer that is evacuated to a certain degree, reducing heat exchange between the inside and outside of the tank. These tanks feature large volumes and high transportation efficiency.

The common vacuum insulation methods include high vacuum insulation, vacuum porous insulation, and high vacuum multilayer insulation. Given the extremely low boiling point and small latent heat of vaporization of liquid helium, which is only one-tenth that of liquid oxygen, its long-term storage is very challenging. Conventional insulation methods, such as vacuum multilayer insulation, are inadequate for liquid helium storage needs [5]. Special insulation methods, such as high vacuum multilayer insulation with a liquid nitrogen shielding screen, are required.

High vacuum multilayer insulation is a method where multiple alternating layers of spacing materials and reflective screens are placed in the interlayer space, which is then evacuated to a high vacuum. Generally speaking, it involves installing radiation screens (like liquid nitrogen cold screens) in the vacuum interlayer to minimize radiation heat transfer between surfaces, providing high vacuum levels and excellent insulation performance. This method is, therefore, widely used in the structural design of liquid helium tank containers.

1. Structure of Liquid Helium Tank Containers

Liquid helium tank containers primarily consist of an inner container, an outer tank, and a liquid nitrogen cold screen. Taking a 40 m^3 tank container as an example, its internal structure is shown in Fig. 7.16.

Liquid helium is stored in the inner container, and the liquid nitrogen cold screen is placed in an intermediate container, with both containers housed within the outer tank. This setup forms two high vacuum spaces between the inner container and the liquid nitrogen cold screen, and between the liquid nitrogen cold screen and the outer tank.

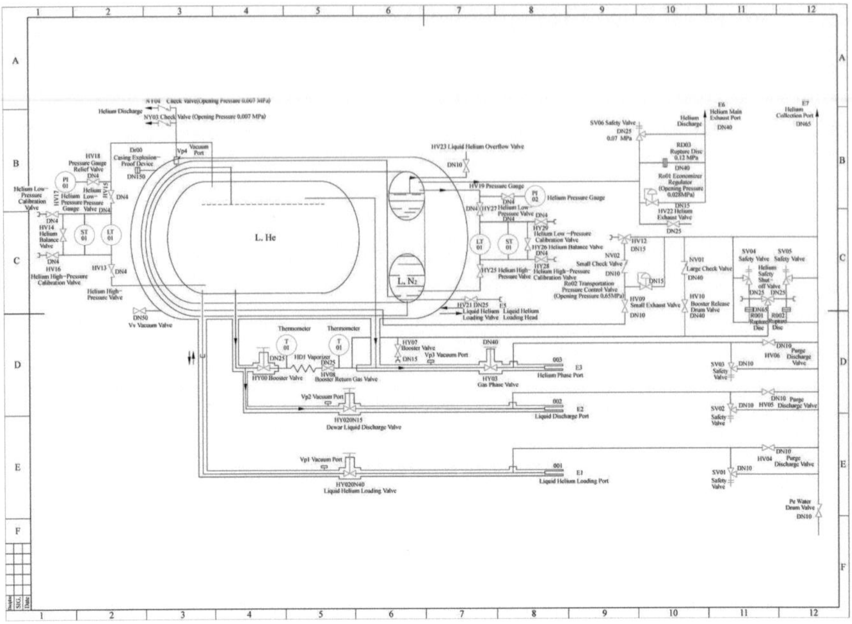

Fig. 7.16 Internal structure of a liquid helium tank container

The liquid nitrogen cold screen acts as a low-temperature protective layer for the inner tank, creating − 196 °C radiative heat wall. This reduces the temperature difference between the walls, minimizes radiative heat flux, blocks external environmental heat radiation, and decreases the evaporation rate of liquid helium.

2. Manufacturing of Liquid Helium Tank Containers

Common volumes for liquid helium tank containers include 20, 40, 60, 80, and 120 m^3. Currently, companies such as Gardner Cryogenics (part of Air Products and Chemicals Inc.) and Linde have the capability to manufacture large-capacity liquid helium tank containers ranging from 40 to 120 m^3. In China, the testing and preliminary application stages are underway. Institutions like the TIPC of the CAS, China Petroleum Engineering and Construction Corporation, Zhangjiagang CIMC Sanctum Cryogenic Equipment Co., Ltd., and Sichuan Air Separation Plant Group are stepping up their development efforts, gradually acquiring the design and manufacturing capabilities for liquid helium tank containers.

The TIPC of the CAS, in collaboration with Aerosun Corporation, has developed a 40 m^3 mobile liquid helium tank container, which has undergone field testing. Zhangjiagang CIMC Sanctum Cryogenic Equipment Co., Ltd. has also developed a 40 m^3 mobile liquid helium tank container. China Petroleum Engineering and Construction Corporation has developed a 120 m^3 fixed liquid helium tank container, which has passed the acceptance test.

3. Technical Characteristics of Liquid Helium Tank Containers

Taking the 40 m^3 liquid helium tank container as an example, Table 7.2 lists the technical parameters of Gardner Cryogenics and Linde.

Table 7.3 lists the materials for the inner and outer shells and the cold screen of liquid helium tank containers from China and some other countries.

4. Usage of Liquid Helium Tank Containers

After being produced, liquid helium can be stored directly in liquid helium tank containers. For first-time use or if contamination is confirmed, the tank should be vacuumized, then purged and replaced with nitrogen, and then with helium, repeatedly until the nitrogen content inside the tank container is $\leq$ 2 mL/m^3 (ppmV), and the oxygen and argon contents are $\leq$ 1 mL/m^3 (ppmV). The tank container is then pre-cooled using liquid nitrogen and liquid helium, and the helium used for precooling can be purified and recovered.

Table 7.2 Technical parameters of **40 m³** liquid helium tank containers [6]

No.	Item	Gardner cryogenics	Linde
1	Storage volume/m^3	40.95	41.00
2	Storage pressure/kPa	7–640	0–600
3	Tank design pressure/kPa	746	–
4	Tank design temperature/°C	– 269 to 37.8	–
5	Liquid nitrogen shield tank volume/m^3	1.66	1.66
6	Liquid nitrogen shield storage pressure/kPa	7–42	0–70
7	Structure	Liquid nitrogen shield multilayer vacuum insulation	Liquid nitrogen shield multilayer vacuum insulation
8	Material	Inner tank: stainless steel; outer shell: carbon steel	Stainless steel throughout
9	Expected helium leak heat loss/(Btu/h)	15	–
10	Maximum helium leak heat loss/(Btu/h)	$\leq$ 22.5	–
11	No-loss transportation time/d	45	40
12	Empty weight/kg	18,400	15,800
13	Fully loaded weight/kg	27,720	23,000
14	External dimensions, mm × mm × mm	–	12,192 × 2438 × 2951

Note The table lists only the known data, which do not indicate superiority or inferiority

Table 7.3 Materials of liquid helium tank containers from china and some other countries

No.	Institution/company	Inner container material	Outer shell material	Liquid nitrogen cold screen material
1	Gardner Cryogenics	SA240-304N	Carbon steel	304
2	The Linde Group	X2CrNiNN18-10	Stainless steel	304
3	Technical Institute of Physics and Chemistry of the Chinese Academy of Sciences	316 L	Carbon steel	304
4	Zhangjiagang CIMC Sanctum Cryogenic Equipment Co., Ltd	316 L	Carbon steel	304
5	China Petroleum Engineering and Construction Corporation	316 L	Carbon steel	304

To minimize leakage, the connections for liquid helium tank containers use insertion interfaces, and both the filling and extraction of liquid helium use vacuum jackets. Typically, the daily evaporation rate of a liquid helium tank container is $\leq 0.3\%$ (when the filling rate is 90%), and the maximum overall leakage rate at design pressure should not exceed 1.0×10^{-6} mbar L/s.

Compared to hydrogen, oxygen, nitrogen, and neon, helium has the highest sensible heat and latent heat ratios. On that account, liquid helium tank containers enable the recovery of sensible heat from helium, thereby reducing heat loss. By dint of structural design, the cooling capacity of helium evaporated from the container can be used to remove heat from the outer wall of the liquid nitrogen cold screen, lowering the wall temperature and reducing heat conduction to the inner container.

During transportation, the liquid nitrogen cold screen of the liquid helium tank container should maintain at least 35 days, and the cold screen charge should not fall below 5% to effectively control the daily evaporation rate. For sea transport, a suspension system is required to ensure the stability of the inner container of the liquid helium tank container during transit. The design acceleration should meet the requirements of the Chinese national standard *Supervision Regulation on Safety Technology for Transportable Pressure Vessel* (TSG R0005-2011).

7.2.2 Liquid Helium Dewar Flask

Named after their inventor, James Dewar, liquid helium Dewar tanks have a smaller volume compared to liquid helium tank containers [7, 8]. When the amount of liquid helium required does not reach the volume of a tank container, Dewar tanks become the preferred choice for storage, transport, and end-use of liquid helium. The main

Fig. 7.17 Liquid helium Dewar tank

specifications currently range from 100 L to 500 L. A diagram of a liquid helium Dewar tank is shown in Fig. 7.17.

1. Structure of Liquid Helium Dewar Tanks

Liquid helium Dewar flasks are high-vacuum insulated containers made of double-layer copper spheres [9]. The inner shell is welded with low-thermal-conductivity silver tubes and then connected to the outer shell, creating a vacuum between them. The lower part of the vacuum interlayer contains a small amount of getter glue or activated carbon to adsorb any trace gases that may have seeped into the vacuum interlayer due to poor workmanship or materials. As there is a vacuum interlayer between the inner and outer shells, a neck tube is installed at the top of the sphere to inject and pour out the cryogenic liquid, while also supporting the sphere. To prevent shaking, elastic pads are used to fix the inner and outer shells.

Although Dewar tanks are small-capacity storage and transportation devices, they are safe, reliable, easy to drain empty, have low static daily evaporation rate, and are convenient for transportation.

2. Manufacture of Liquid Helium Dewar Tanks

The production technology for liquid helium Dewar tanks is relatively simple and has mature applications. Wessington Cryogenics in the UK can produce liquid helium Dewar flasks with capacities over 500 L, and Chart Industries in the US manufactures Dewar flasks ranging from 100 to 500 L.

In 1988, Shanghai Jiao Tong University in China developed a transportable liquid helium container (TLHE-100 type), capable of short-distance transport of liquid helium and conventional ultra-low temperature experiments, marking the first reliable liquid helium transport container in China. In recent years, with progress in science and technology, the manufacturing and application of liquid helium Dewar tanks in China have also progressed. Companies such as Beijing Sinoscience Fullcryo Technology Co., Ltd. and Chart Cryogenic Engineering Systems (Changzhou) Co., Ltd. can produce a range of liquid helium Dewar flasks with various capacities.

3. Characteristics of Liquid Helium Dewar Flasks

Using a 100-L liquid helium Dewar tank as an example, Table 7.4 lists the technical parameters of the Dewar tanks from Beijing Sinoscience Fullcryo Technology Co., Ltd. and Chart Cryogenic Engineering Systems (Changzhou) Co., Ltd.

Liquid helium Dewar tanks are made from austenitic stainless steel. The large neck size allows for the insertion of standard helium lines, which can eliminate thermal oscillations. Overall, the structural design of Dewar flasks minimizes the loss of liquid helium during storage, transportation, and distribution. They incorporate high-efficiency vacuum insulation technology, vapor-cooled multi-cold-screen technology, and low-temperature condensation adsorption high vacuum technology, all of which contribute to a decrease in the evaporation rate of liquid helium.

4. Use of Liquid Helium Dewar Tanks

Similar to liquid helium tank containers, Dewar tanks should be vacuumized, purged and replaced with nitrogen, and then with helium several times to reduce impurity levels to acceptable ranges before first use or after contamination. The tanks are

Table 7.4 Technical parameters of **100 L** liquid helium Dewar tanks

No.	Item	Beijing Sinoscience Fullcryo Technology Co., Ltd.	Chart Cryogenic Engineering Systems (Changzhou) Co., Ltd.
1	Effective volume/L	100	100
2	Daily static evaporation rate/%	1.00	1.25
3	Working pressure/MPa	0.07	–
4	Design pressure/MPa	0.09	0.08
5	Diameter/mm	650	606
6	Height/mm	1500	1546

Note The data presented is not a strict technical parameter comparison, but a listing of available data. For precise details, please consult the manufacturers pre-cooled with liquid nitrogen and liquid helium before being used for storage. The helium used for precooling can be recovered after purification.

Due to their small volume and low pressure, liquid helium Dewar tanks are suitable for transport by air or road. The tanks are typically equipped with liquid valves, gas valves, vent valves, and pressurization valves. When used solely for storing liquid helium, the pressurization valve must be tightly closed to prevent leakage. When gaseous helium is required at the end-use point, the pressurization valve can be opened, allowing the liquid to vaporize and supply gas. The pressure gauge should be monitored to determine if there is a liquid phase inside the Dewar tank. During use and transport, Dewar tanks should be kept upright to avoid tilting.

7.2.3 Liquid Helium Loading

The primary storage and transportation equipment for liquid helium are liquid helium tank containers and liquid helium Dewar tanks. Dewar tanks, due to their relatively small size, have control valve assemblies and fittings installed on the top and can be connected to various standard-sized hoses. This makes them flexible and convenient to operate, requiring no specialized loading equipment, and suitable for road, air, and water transport. On the other hand, liquid helium tank containers, given their larger volume, are mainly transported by road and water. The choice of transport depends on the distance and other factors. Regardless of the method, the process of loading liquid helium into tank containers is always necessary.

In industrial facilities, liquid helium storage equipment is typically placed on elevated platforms, with mobile liquid helium tank containers positioned below. When loading is required, the transfer is controlled via loading valve boxes, utilizing gravity to move the liquid helium from the storage equipment on the platform to the mobile tank containers on the tube trailers. The loading process is primarily controlled by monitoring the liquid level, and loading is stopped once the designated level is reached. Liquid helium tank container loading often uses weighing for transaction settlement.

The pipeline system of liquid helium tank containers is relatively complex. Before transport, it is essential to check the safety of the liquid nitrogen cold screen to ensure it meets the requirements for the transport distance and duration. In case of a leak during transport, the tank should be promptly moved to a sparsely populated, open, and safe place. During the release, both personnel and vehicles should be positioned upwind, and the release should be conducted gradually, with safety personnel present for operation and supervision.

7.3 Selection of Storage and Transport Methods

As previously mentioned, there are various methods for storing and transporting helium. Gaseous helium can be stored in tanks, high-pressure gas cylinder bundles, bundle containers, gas cylinders, gas holders, helium storage reservoirs, and so forth, while liquid helium can be stored in liquid helium tank containers and Dewar tanks. The appropriate storage and transport method should be selected based on factors such as the scale of transportation and the distance.

7.3.1 Selection of Gaseous Helium Storage and Transportation Equipment

The recommended selection of gaseous helium storage and transportation equipment is shown in Table 7.5.

From Table 7.5, it can be seen that gas holders, due to their low pressure and small volume, are only suitable for small-scale residual gas storage. Gas cylinders, despite their relatively small volume, are flexible and convenient, typically used for small-scale gaseous helium storage and transport. Storage tanks, as high-pressure equipment, can be customized to different volumes as needed, suitable for larger-scale storage. Gas cylinder bundles, also high-pressure equipment, can increase storage volume by adding more bundles, making them flexible for both fixed storage and mobile transport.

7.3.2 Selection of Liquid Helium Storage and Transportation Equipment

Currently, the most commonly used liquid helium storage and transportation equipment includes liquid helium Dewar tanks and liquid helium tank containers. Dewar tanks are mainly used for small-scale storage and transport. Liquid helium tank containers, available in sizes such as 20, 40, 60, 80, and 120 m^3, are suitable for large-scale liquid helium storage and transport. The 40 m^3 liquid helium tank container is the most widely used model, manufactured according to ISO (International Organization for Standardization) standards, and is, therefore, also known as an ISO tank container.

Table 7.5 Selection of gaseous helium storage and transportation equipment

–	Storage tanks	High-pressure gas cylinder bundle	Gas cylinder	Gas holder
Material	Q345R (normalized)	4130X	Quality carbon manganese steel	PVC
Manufacturing standard	*Pressure Vessels* (GB/T 150.1 –150.4-2011)	*Regulation on Safety Technology for Gas Cylinder* (TSG 23-2021)	*Regulation on Safety Technology for Gas Cylinder* (TSG 23-2021)	–
Manufacturing process	Multi-layer wrapping	Large-volume seamless steel gas cylinders	Seamless gas cylinders	–
Operating pressure/MPaG	20–25	20	15	Approximately 0.004
Gas tightness test pressure/MPaG	–	20	15	–
Hydraulic test pressure/MPaG	34.38	33.4	22.5	–
Safety facilities	Safety valve on outlet pipe	Safety valve on outlet pipe	Safety valve on outlet pipe	Pressure relief and water seal cylinders
Advantages	High pressure resistance	Large volume, small footprint	Portable, easy to maintain	Low cost
Disadvantages	Complex manufacturing	–	Small filling volume	Very small filling volume, only suitable for low-pressure residual gas recovery
Applicable conditions	Large volume, high pressure	Large volume, high pressure	Small volume, medium pressure	Small volume, low pressure
Transportability	No	Yes	Yes	No

7.3.3 Comparison of Gaseous Helium and Liquid Helium Storage and Transport

To increase storage volume and efficiency, gaseous helium is often stored and transported under high pressure. The advantages of this method include a relatively simple process and lower energy consumption, particularly by reducing the need for helium liquefaction equipment. However, due to the lower storage density of gaseous helium compared to liquid helium, the frequency of transport is higher, and as the scale of storage and transport increases, transportation costs rise rapidly. Therefore, the transportation efficiency of high-pressure storage is relatively low.

In contrast, liquid helium storage and transport offer the advantage of high storage density, effectively reducing transport frequency and costs, making it suitable for large-scale, long-distance transportation. Nevertheless, this method involves complex cryogenic refrigeration processes, longer equipment procurement cycles, and higher energy consumption.

Based on an annual helium production scale of 1×10^6 m^3 and a transportation distance of 3000 km, a comparison of investment and costs between gaseous helium and liquid helium storage and transport methods is conducted. The cost calculation for both methods excludes the expenses for natural gas purification at the front end and helium refinement and only includes the major transportation costs without considering road conditions and other factors. The gaseous helium plan includes 2 diaphragm compressors, each costing RMB 2 million, and 24 multi-layer wrapped storage tanks of 17 m^3 each for helium storage, with tube trailers priced at RMB 1.1 million per unit. The liquid helium plan includes a helium liquefier costing RMB 20 million, liquid helium tank containers for storage, and specialized liquid helium tank trucks priced at RMB 8.5 million per unit.

Both gaseous and liquid helium plans are calculated based on a bi-monthly turnover. Production costs primarily include operational electricity costs, production equipment depreciation, and management fees. Electricity costs are calculated at RMB 0.6/kWh, production equipment depreciation is based on a 10-year period, and management fees are considered as 50% of the depreciation costs. Transportation costs mainly include vehicle depreciation, transportation expenses, and vehicle maintenance and inspection. Vehicle depreciation is based on a 10-year period, transportation expenses are calculated at RMB 50,000 per trip, with inspection fees of RMB 2,000 per trip for gaseous helium and RMB 5000 per trip for liquid helium. The economic comparison of the two plans is shown in Table 7.6.

The unit total cost of the gaseous and liquid helium plans consists of unit production cost and transport cost. With a fixed transport scale, changes in transport distance do not affect the unit production cost of both plans. However, the unit transport cost of gaseous helium increases more rapidly with distance, making gaseous helium suitable for short-distance transport. From Table 7.6, it can be seen that while the production cost of the gaseous helium plan is lower, its transport cost is higher. Conversely, the liquid helium plan has higher production costs but lower transport costs, resulting in a lower unit total cost for liquid helium. Therefore, under the same input conditions, the liquid helium plan is more suitable for long-distance, large-scale transportation.

Table 7.6 Economic comparison of gaseous helium and liquid helium plans [10]

No.	Parameter	Gaseous helium plan	Liquid helium plan
1	Production scale/($10^3 m^3$/a)	1000	1000
2	Transport distance/km	3000	3000
3	Transport trips/(trips/a)	250	41
4	Number of transport vehicles	42	8
5	Production equipment investment/¥0,000	400	2000
6	Factory storage equipment investment/¥0,000	3600	2200
7	Transport vehicle investment/¥0,000	4620	6800
8	Production cost/¥0,000	848.76	1107.50
9	Transport cost/¥0,000	1725.2	804.0
10	Unit production cost/(RMB/m^3)	8.4876	11.0750
11	Unit transport cost/(RMB/m^3)	17.25	8.04
12	Unit total cost/(RMB/m^3)	25.74	19.12

References

1. Zhao LH, Kang TR. Introduction of design of multilayer-wrapped high-pressure vessels. Mech Manag Dev. 2013;5:50–52 (in Chinese).
2. Liu YQ, Xiao ZG, Fan AL. Advantages of multi-layer wrapped structures in high-pressure vessels. Technol Dev Enterpr. 2013;32(5):38–39 (in Chinese).
3. Tang L, Ding GS, Han CA. Underground helium storage. Beijing: Petroleum Industry Press; 2020 (in Chinese).
4. Hamacher T. Wasserstoff als strategischer Sekundärenergieträger. Wasserstoff und Brennstoffzelle: Technologien und Marktperspektiven. 2013;2017:1–23.
5. Zhang P, Wang RZ. Heat transfer of superfluid helium. Beijing: Science Press; 2009 (in Chinese).
6. Li JF, Zhang RC, Chen JG. Current situation and key technology of liquid helium storage tank. Low Temp Spec Gas. 2021;39(5):8–10 (in Chinese).
7. Dhurjad S, Mahindrakar A, Mache A, et al. Design and development of liquid helium storage dewar. In: Biennial international conference on future learning aspects of mechanical engineering. Singapore: Springer Nature Singapore; 2022. pp. 223–242.
8. Pandharpatte P, Kurkure K, Kore S, et al. Review and development of thermal design of a cryogenic dewar. In: Biennial international conference on future learning aspects of mechanical engineering. Singapore: Springer Nature Singapore; 2022. pp. 319–331.
9. Heidt C, Grohmann S, Süßer M. Modeling the pressure increase in liquid helium cryostats after failure of the insulating vacuum. AIP Conf Proc. 2014;1573(1):1574–1580.
10. Zhang RC, Li JF. The technical and economic comparison of storage and transportation with liquid helium and high-pressure gas helium. Low Temp Spec Gas. 2019;37(6):17–21 (in Chinese).

Chapter 8
Equipment for Helium Extraction from Natural Gas

Zhiming Jiang and Yingjie Xiong

The process for helium extraction from natural gas is complex and involves a variety of equipment. In addition to conventional equipment such as shell-and-tube heat exchangers, filters, and pumps, specialized devices like plate-fin heat exchangers, helium concentration columns, gas separation membranes, compressors, and helium turboexpanders are also utilized. The performance of these devices significantly impacts the operation of the helium extraction process from natural gas.

Among the various equipment for helium extraction from natural gas, the plate-fin heat exchanger is a key device for multi-stream heat exchange. The helium concentration column leverages the difference in boiling points of various components at low temperatures to concentrate helium. Gas separation membranes utilize the varying permeation rates of gas molecules through membrane materials to achieve helium concentration. Various compressors are applied in multiple process units, and helium turboexpanders use the adiabatic isentropic expansion of the medium to achieve low temperatures. This chapter provides an introduction to the key equipment used in the helium extraction process.

8.1 Plate-Fin Heat Exchanger

Plate-fin heat exchangers, commonly known as aluminum plate-fin heat exchangers, were initially used in the aerospace industry and later introduced to the field of cryogenics. They are now widely used in refrigeration applications [1]. The traditional manufacturing process for plate-fin heat exchangers involves salt-bath dip

Z. Jiang (✉) · Y. Xiong
China Petroleum Engineering & Construction Corporation Southwest Company, Chengdu, China
e-mail: jiangzhim_sw@cnpc.com.cn

Y. Xiong
e-mail: xiongyingjie_sw@cnpc.com.cn

L. Pu (ed.), *Helium Engineering Technologies*,
https://doi.org/10.1007/978-981-96-4086-7_8

brazing, which exploits the large heat capacity and rapid heat conduction of the dip to achieve quick and uniform welding, resulting in high production efficiency. However, this approach leaves corrosive chloride salts on the welded parts, necessitating complex post-welding treatments. Currently, vacuum brazing technology is more commonly employed. This method, performed under vacuum conditions, requires no flux and enables efficient brazing with high welding quality, high yield rates, and minimal environmental pollution. It effectively addresses the corrosion issues caused by chloride residues from salt-bath dip brazing [2]. Plate-fin heat exchangers have a broad application range, suitable for evaporators, condensers, and low- and ultra-low-temperature scenarios. Due to the presence of fins on both sides of the heat exchange plates, they are particularly suitable for scenarios where the heat transfer coefficients on both sides are similar.

In natural gas cryogenic helium extraction units, the natural gas is gradually cooled through multiple heat exchanges until the majority of methane and other hydrocarbon components are liquefied. During this heat exchange process, the operating temperature is often lower than the ambient temperature. To minimize cooling losses, heat exchangers operating at lower temperatures are housed within a well-insulated enclosure known as the helium extraction cold box. It is a critical component of equipment in natural gas cryogenic helium extraction.

8.1.1 Structure

A plate-fin heat exchanger is consists of several components, including the core, head, connecting pipes, supports, etc. Among them, the core serves as the heart of the plate-fin heat exchanger, forming a complete unit once welded with heads and supports, as illustrated in Fig. 8.1.

The core is composed of multiple layers (channels) stacked in a specific alternating sequence of hot and cold streams, which are joined by vacuum brazing. The core can accommodate up to 15 fluid streams, with each stream comprising a certain number of layers (channels). The layers (channels) consist of fins, partition plates, seal strips, and guide plates.

Heat transfer and exchange in a plate-fin heat exchanger occur within the core. As the hot fluid flows through the channel, heat is transferred to the hot-side fins via convection. The heat is then conducted through the partition plate to the fins in the cold channel, where the cold fluid removes the heat by convection as it flows through.

8.1.2 Features

The worldwide manufacturing standards for plate-fin heat exchangers are relatively advanced. The current Chinese industry standard, *Aluminum Plate-Fin Heat*

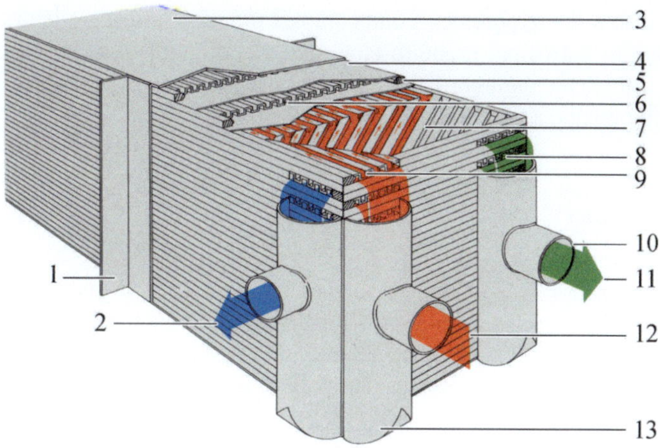

1 - Support Angle; 2 - Stream A; 3 - Outer Plate; 4 - Partition Plate; 5 - Bar; 6 - Heat Transfer Fins; 7 - Tunning Distribution Fins; 8 - Inlet and Outlet Ports; 9 - Port Distribution Fins; 10 - Nozzle; 11 - Channel B; 12 - Stream C; 13 - Header

Fig. 8.1 Structure of plate-fin heat exchanger [3]

Exchangers (NB/T 47006-2019), stipulates that the design pressure limit for aluminum plate-fin heat exchangers has reached 10.0 MPa, consistent with the ALPEMA 2010 standards [4]. The main features of plate-fin heat exchangers include [5]:

(1) High heat transfer efficiency. The turbulence created by the fins disrupts the fluid boundary layer, thereby increasing the heat transfer coefficient. The thin fins and partition plates have low thermal resistance, which contributes to enhanced high heat transfer efficiency.
(2) Compact structure. The secondary surface of the fins can be significantly expanded, with specific surface areas reaching 1000–1500 m^2/m^3. New low-pressure fin products can achieve a heat transfer area of up to 1900 m^2/m^3, leading to a more compact equipment structure due to the large heat transfer area within a unit volume.
(3) Lightweight. Predominantly made of aluminum alloy, plate-fin heat exchangers are lightweight, with their weight being only 10%–65% that of shell-and-tube heat exchangers under similar heat transfer conditions.
(4) Strong adaptability. Plate-fin heat exchangers can be used for heat exchange between different phases of fluids, such as liquid–liquid, gas–liquid, and gas–gas. Adjustments to the flow path arrangements and combinations can help accommodate cross-flow, counter-flow, and multi-stream heat exchange conditions. These exchangers can be configured in parallel, series, or parallel-series configurations to meet the heat exchange needs of large-scale equipment, enabling mass production and standardization in industrial applications, thereby reducing costs.

(5) Strict manufacturing requirements and complex process control.
(6) Not resistant to corrosion, prone to clogging, and difficult to maintain. Suitable only for clean, non-corrosive, non-deposit-forming, non-scaling, and non-clogging media.

8.1.3 Applications

Plate-fin heat exchangers are widely used in air separation, natural gas processing, ammonia synthesis, aerospace, automotive, refrigeration, and air conditioning. In helium extraction units from natural gas, aluminum plate-fin heat exchangers are primarily used for cryogenic processing, particularly in 2.5 kW and 4.5 K helium refrigeration systems.

China's first application of plate-fin heat exchangers for extracting helium from natural gas occurred at the Rongxian Helium Extraction Unit, with a natural gas processing capacity of 4×10^5 m^3/d. The cold box was 15 m × 3.5 m × 3.2 m in dimensions, recording a crude helium concentration of 68.3% exiting the cold box. By the end of 2023, the processing capacity of single-column helium extraction units from natural gas with plate-fin heat exchangers in China has exceeded 1×10^6 m^3/d. Table 8.1 presents the application status of plate-fin heat exchangers in typical helium extraction plants in China.

8.2 Helium Concentration Column

The helium concentration column is a crucial component of cryogenic rectification equipment used in the process for helium extraction from natural gas. Depending on the specific cryogenic rectification process, either a single-column or dual-column system may be used. During the cryogenic extraction of helium from natural gas, hydrocarbons such as methane and the majority of nitrogen liquefy at low temperatures and undergo rectification separation within the helium concentration column.

Table 8.1 Application of plate-fin heat exchangers in typical helium extraction plants in China

Equipment	Dimensions/(mm × mm × mm)	Heat load/kW	Number of streams
Natural gas liquefaction heat exchanger	7000 × 1200 × 1400	10,600	9
Demethanization heat exchanger	4800 × 300 × 380	250	6
Feed gas heat exchanger	5100 × 900 × 900	4200	6
Cryogenic heat exchanger	6000 × 500 × 560	360	5

Lighter components such as helium, hydrogen, and neon are separated at the top of the column, resulting in crude helium.

8.2.1 Structure

Helium concentration columns can be categorized into plate columns and packed columns based on their internal structure.

1. Plate Column

A plate column is a device used for gas–liquid contact mass transfer across multiple stages, consisting of a cylindrical tower body with several trays installed at specified intervals. In a plate column, the gas phase contacts the liquid phase on the trays through bubbling, enabling the transfer of substances. This process allows lower-boiling-point gases such as helium and hydrogen to accumulate at the top of the column for the purpose of concentration.

The main components of a plate column include the tower shell, trays, demister, and tower bottom support. Taking the bubble tower as an example, its typical structure is illustrated in Fig. 8.2. The trays are perforated with certain vent holes covered by round caps, with serrated edges or small holes at the lower edge of the caps. During operation, the liquid first enters the top tray, flows across the tray surface from one end to the other, overflows the outlet weir, and reaches the next tray through the downcomer, cascading downwards tray by tray. The overflow weir ensures a certain liquid level on the tray, submerging the small holes (or serrations) under the bubble caps. Gas enters from beneath the lowest tray and flows upward through the vent pipes on the trays. Due to the presence of liquid on the trays, the gas is dispersed into bubbles as it passes through each tray via the vent pipes. The gas load should ensure that the bubbles come into close contact, continuously merge and rupture, transforming the liquid layer into a foam layer (continuous liquid phase). Upon leaving the foam layer, some liquid is carried out as mist, dispersed between the liquid surface and the upper tray (continuous gas phase). The interaction between the bubbles and liquid mist increases the contact area between the two phases and induces turbulence to a certain degree, thereby enhancing mass transfer rates. The gas and liquid phases flow in cross-current on each tray and counter-current throughout the column [6].

Plate columns offer advantages such as high production capacity, efficient mass transfer, operational flexibility, low cost, and ease of scale-up, but their complex structure results in high pressure drop and large liquid holdup [7].

2. Packed Column

A packed column enhances mass transfer through counter-current contact between gas and liquid phases within a packed bed. Lower-boiling-point components such as helium and hydrogen are concentrated at the top of the column.

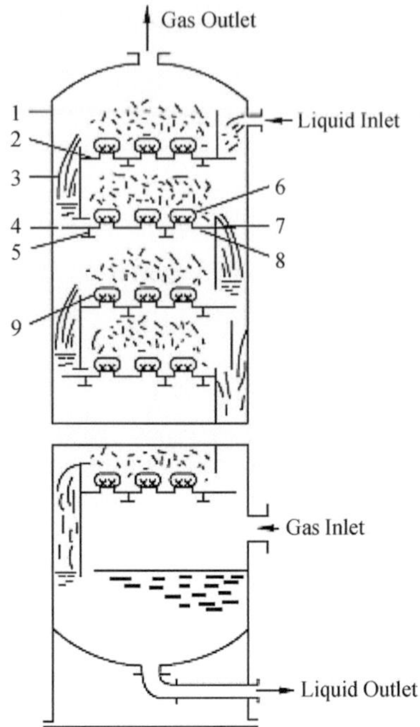

1 - Column Shell; 2 - Tray; 3 - Downcomer; 4 - Support Ring; 5 - Reinforcement Beam; 6 - Bubble Cap; 7 - Overflow Weir; 8 - Weep Hole; 9 - Upriser

Fig. 8.2 Typical structure of a plate column

The main components of a packed column include the tower shell, packing material, packing support, bed limiters, and liquid redistributors. A typical structure is shown in Fig. 8.3. A certain height of packing is filled inside the shell of a vertical cylindrical tower, with the packing placed on a support plate near the column bottom. Gas enters from the bottom inlet, while liquid is introduced from the top, where it is dispersed by a distributor onto the packing surface. As the liquid flows through the packing, it tends to move towards the column wall. Therefore, when the packing height is high, it needs to be handled in sections, between which liquid redistributors are installed. The liquid flows down through the packing gaps, spreading as a film or droplets on the packing surface, which provides a contact surface for gas–liquid mass transfer. The gas and liquid phases interact countercurrently within the packing bed, with their compositions continuously changing along the vertical axis [6].

Packed columns, using packing as the gas–liquid contact element, allow the gas and liquid phases to undergo continuous counter-current contact within the packing bed. They have a simple structure, low pressure drop, high efficiency, small diameter, and low cost, but they also exhibit significant scale-up effects and pose challenges in installation and maintenance.

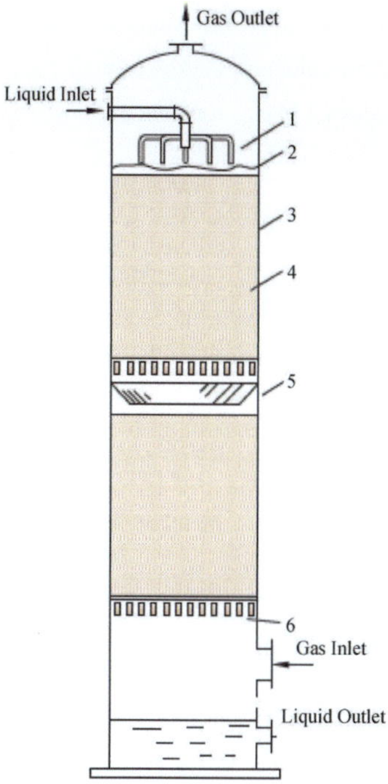

1 - Liquid Distributor; 2 - Packing Retainer; 3 - Column Shell; 4 - Random Packing; 5 - Liquid Redistributor; 6 - Packing Support

Fig. 8.3 Typical structure of a packed column

8.2.2 *Features*

The features of plate columns and packed columns are shown in Table 8.2.

The choice of helium concentration columns for extracting helium from natural gas depends on factors such as the properties of the medium, gas–liquid ratio, and operating pressure. Packed helium concentration columns use packing as the gas–liquid contact element, with the two phases continuously contacting countercurrently in the packing layer, offering advantages such as simple structure, low pressure drop, high efficiency, small diameter, and low cost. In small-scale natural gas cryogenic helium extraction systems, given the liquid load being relatively small, packed columns with random packing are commonly used. For large-scale helium extraction systems from natural gas, plate columns may be applied.

Additionally, the cryogenic helium concentration system used in natural gas extraction is a complex system comprising rectification columns and heat exchangers.

In the process of extracting helium from natural gas, due to the operating temperature as low as − 180 °C and the high insulation requirements, minimizing cooling loss during the process becomes a key consideration [9]. With progress in mechanical manufacturing technology, heat exchangers can be integrated with concentration

Table 8.2 Features of plate columns and packed columns [7, 8]

No.	Item	Plate column	Packed column
1	Pressure drop per theoretical plate	About 1 kPa	Random packing: ~ 0.3 kPa; structured packing: ~ 0.15 kPa
2	Separation efficiency (HETP)	Relatively stable, higher efficiency with larger diameters	Packed columns with conventional structured packing are less efficient than plate columns; the efficiency of packed columns with new random packing is comparable to that of plate columns
3	Liquid–gas ratio	Wide applicable range	Certain requirements for liquid spray rates
4	Liquid holdup	Relatively large	Relatively small
5	Material	Typically metal	Metal or non-metal corrosion-resistant materials
6	Installation and maintenance	Relatively easy	Relatively difficult
7	Cost	Generally lower than packed columns for larger diameters	Lower than plate columns for DN800 and below; cost increases significantly with diameter

columns within a cryogenic insulated shell. Additionally, the condenser at the top of the column and the reboiler at the bottom can be integrated into the concentration column, in a bid to shorten the length of external pipelines and effectively reduce the system's cooling loss.

8.2.3 Applications

A typical natural gas helium extraction project adopts a co-production process that combines natural gas liquefaction and helium extraction. This process integrates both primary and secondary concentration columns within the same cold box, along with cryogenic heat exchangers. This design reduces the length of external cryogenic pipelines and minimizes the system's cold energy loss, as shown in Fig. 8.4.

Fig. 8.4 Typical dual-column cryogenic plant for helium extraction from natural gas

8.3 Gas Separation Membrane

Gas separation membranes can be applied to various natural gas processing stages, including acid gas removal, dehydration, nitrogen separation, and heavy hydrocarbon separation [10]. In the 1950s, some countries began researching membrane separation of helium, but commercial application was hindered by such as poor selectivity, low permeability, and high costs. It was not until 1965, when Stern proposed using polytetrafluoroethylene (Teflon FEP)—a copolymer of tetrafluoroethylene and hexafluoropropylene—for selective helium recovery from natural gas, that membrane separation technology was shown to have potential for large-scale helium separation from natural gas [11, 12]. In recent years, progress in the membrane industry has greatly boosted membrane selectivity and permeability, making gas separation membranes commercially viable for hydrogen and helium separation.

Due to the structural characteristics of separation membranes, the capacity of a single membrane is limited, often requiring multiple membrane units to be used in parallel and mounted on skids. A membrane separation skid typically comprises membrane modules, piping, valves, and a skid base, with the membrane modules being at the core. Based on the material, membrane modules can be categorized into organic and inorganic membrane separation modules [13, 14].

8.3.1 Types of Membrane Modules

Membrane separation modules can be categorized by their configuration into three types: hollow fiber, spiral-wound, and plate-and-frame modules. Hollow fiber membrane modules are made from hollow fiber membranes, resembling a shell-and-tube heat exchanger, with tens of thousands of hollow fiber strands bundled together and encased in a steel shell, as shown in Fig. 8.5 [15]. They offer such advantages as high packing density (1600–3000 m^2/m^3), low manufacturing cost per unit membrane area, and good pressure stability, yet they cannot be replaced if damaged.

Spiral-wound and plate-and-frame membrane modules are made from flat sheet membranes. A spiral-wound membrane module mainly consists of the membrane, feed channels, permeate channels, spacers, and an outer casing, as shown in Fig. 8.6 [16]. Advantages of spiral-wound membrane modules include simple structure, low cost, and relatively high packing density (< 1000 m^2/m^3). The main disadvantage is that they are difficult to clean [17].

The basic components of plate-and-frame membrane modules include flat sheet membranes, support plates, and spacer plates, as shown in Fig. 8.7. Advantages of plate-and-frame membrane modules include flexible operation, simple assembly, robust structure, as well as ease of maintenance, cleaning, and replacement. However, their packing density is relatively low (100–400 m^2/m^3) [18] (Figs. 8.6 and 8.7).

Among the three types of membrane modules, hollow fiber membrane modules boast the highest packing density with relatively low cost, making them the preferred choice for helium extraction from natural gas.

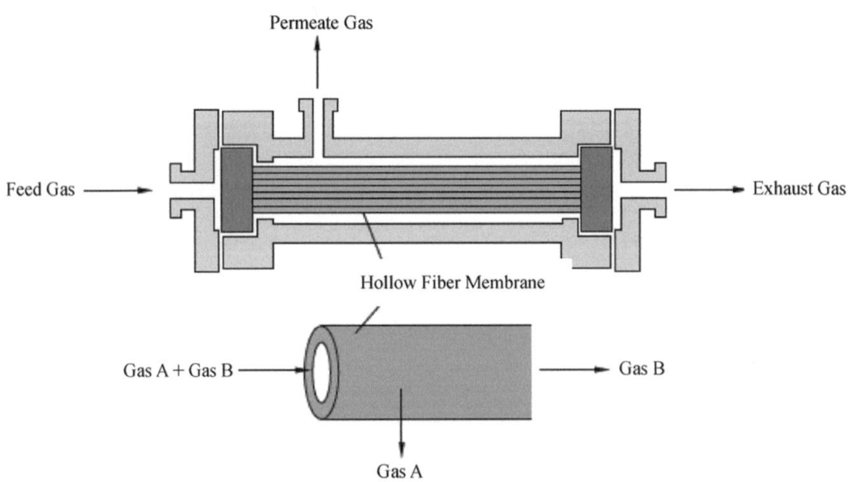

Fig. 8.5 Hollow fiber membrane module [15]

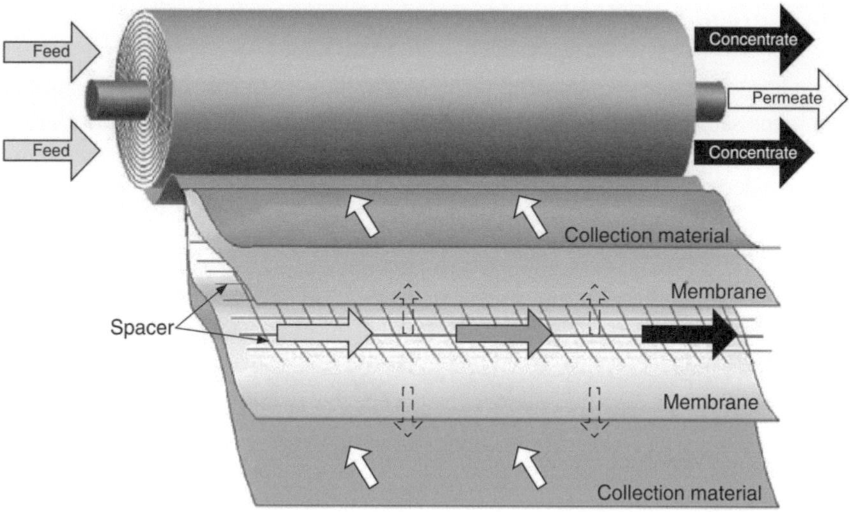

Fig. 8.6 Spiral-wound membrane module [16]

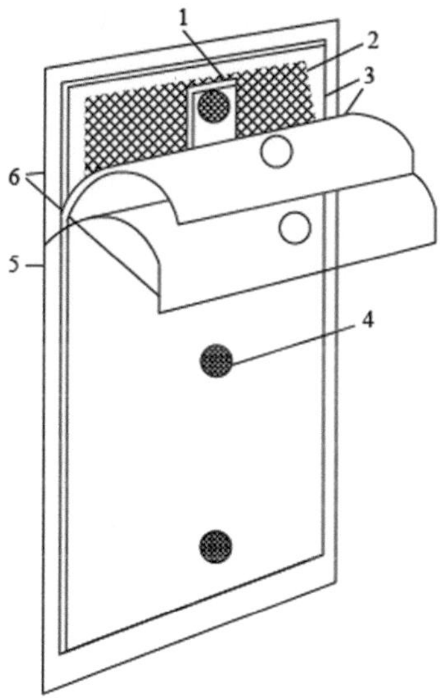

1 - Metal Spacer; 2 - Net; 3 - Paper; 4 - Permeate Gas Outlet; 5 - Heat Seal; 6 - Permeate Membrane

Fig. 8.7 Plate-and-frame membrane module [18]

8.3.2 Membrane Module Materials

The selection of membrane module materials significantly affects the efficiency of membrane separation systems, the lifespan of membrane modules, and the fabrication complexity. Both inorganic and organic membranes can be used in membrane modules, depending on the conditions of the feed gas.

1. Organic Membrane Materials

Organic membranes are advantageous due to their low production costs, good flexibility, and ease of commercialization. Commonly used organic membrane materials in membrane module fabrication include cellulose acetate membrane, polycarbonate membrane, polymethyl methacrylate membrane, and polyimide membrane [19].

(1) Cellulose Acetate Membrane

Cellulose acetate membrane was the first organic membrane material used for helium separation. Gantzel et al. prepared a 41 μm-thick "dense" cellulose acetate membrane by dint of acetone solution evaporation and used it for mixed gas separation studies, indicating that at room temperature, the helium permeation rate was 4.67×10^{-8} mol/($m^2 \cdot s \cdot Pa$), and the selectivity for He/N_2 and He/CH_4 reached 97%, comparable to Teflon FEP. Kumazawa et al. fabricated a hollow fiber membrane from triacetate cellulose for separating helium from air and helium mixtures, achieving a helium recovery rate of 99.9%. They also studied the effects of temperature and inlet pressure on helium permeation rate, finding that inlet pressure had no significant effect at the same temperature, but the permeation rate increased with temperature at constant pressure [20–22].

Cellulose acetate membranes are widely used in ultrafiltration, reverse osmosis, and nanofiltration operations given their low cost. However, their poor acid and alkali resistance, low stability, a typical operating temperature limit of 80 °C, and difficulties in quality control during manufacturing pose challenges for large-scale industrial production.

(2) Polycarbonate Membrane

Polycarbonate membranes offer high strength and chemical stability, making them suitable for helium separation. López-González et al. used hot-press molding at 300 °C to prepare polycarbonate separation membranes and studied the permeation rate of helium through the polycarbonate membrane, as well as the effects of temperature and pressure on the helium separation performance. The results showed that at 0.023 MPa and 30 °C, the helium permeation rate was 2.047×10^{-8} mol/($m^2 \cdot s \cdot Pa$), with permeation rates increasing with temperature but not significantly affected by inlet pressure [23].

However, polycarbonate membranes are brittle, prone to aging and cracking at high temperatures, and entail high manufacturing costs. Despite extensive research, their application in membrane production remains limited.

(3) Polymethyl Methacrylate (PMMA) Membrane

PMMA membrane is a non-crystalline linear thermoplastic material known for its high mechanical strength and good heat resistance. Min et al. studied the effect of PMMA membranes with four different stereoregularities on the separation performance of He, H_2, O_2, N_2, CH_4, and CO_2 mixtures. At near-atmospheric pressure and 35 °C, the helium permeation rate was (1.33–3.33) × 10^{-9} mol/(m^2·s·Pa), with He/CH_4 selectivity ranging from 1300 to 3800, and He/H_2 selectivity from 2.0 to 2.9 [24]. PMMA membranes exhibit excellent performance in separating He/CH_4. The main approach to boost their gas separation performance is through blending and copolymerization modifications. However, currently, PMMA membranes have not yet been applied in industrial production.

(4) Polyimide Membrane

Polyimide is a glassy polymer with strong corrosion resistance, high glass transition temperature, heat resistance, stability, and selectivity, making it an ideal gas separation membrane material. Polyimide helium separation membranes typically have low permeation rates, which can be increased by raising the temperature, although this may reduce gas selectivity. Li Junfang et al. prepared a He/CH_4 mixture with 0.3% helium and used a polyimide hollow fiber membrane for separation experiments. They examined the effects of temperature and pressure, among other conditions, on helium concentration and recovery, finding that after the mixture passed through the polyimide membrane, the helium concentration increased 9–15 times, with recovery rates of 70%–90% [25].

Polyimide membranes exhibit excellent gas separation performance and have been commercially applied in helium extraction from natural gas. Commercial helium extraction membrane modules produced by Air Liquide from France, Evonik Industries AG from Germany, and Air Products and Chemicals Inc. from USA all use polyimide materials.

(5) Comparison of Organic Membrane Materials

Organic membranes for helium separation often cannot simultaneously achieve both high permeability and high selectivity. Membrane performance can be enhanced through various modification techniques. The characteristics of different organic membrane materials are compared in Table 8.3.

2. Inorganic Membrane Materials

Common inorganic membrane materials include silicon dioxide, carbon molecular sieve membranes, and zeolite molecular sieve membranes.

(1) Silicon Dioxide

Table 8.3 Comparison of different organic membrane materials

Membrane types	Advantages	Disadvantages
Cellulose acetate membrane	High helium permeability, strong selectivity	Poor biological stability, weak acid and alkali resistance, difficult quality control in large-scale production
Polycarbonate membrane	High mechanical strength, good chemical stability	Brittle, prone to aging and cracking at high temperatures, high manufacturing cost
Polymethyl methacrylate membrane	High mechanical strength, good thermal stability	Poor separation performance
Polyimide membrane	Strong corrosion resistance, high glass transition temperature, heat resistance, good stability and selectivity	High manufacturing cost, high water absorption, thermal expansion due to water absorption

Silicon dioxide is one of the earliest materials used for helium separation due to its excellent chemical and thermal stability. Silicon dioxide membranes can separate high-purity helium at room temperature with low permeation flux. In practical applications, the permeation flux of helium molecules can be enhanced by raising the temperature. Shelekhin et al. used microporous silicon dioxide fiber membranes to study the permeation rates of He, H_2, CO_2, O_2, N_2, and CH_4 at different temperatures. The results showed that the permeation rate of gas molecules decreased with increasing molecular diameter, with He/CH_4 permeation selectivity exceeding 10,000 at 30 °C. However, the selectivity decreased with increasing temperature [26].

(2) Carbon Molecular Sieve Membranes

Carbon molecular sieve membranes are a type of porous carbon-based inorganic membrane with uniform pore size distribution close to molecular dimensions, offering high gas permeability, good selectivity, strong chemical resistance, and thermal stability. They are usually prepared by pyrolysis and carbonization of polymer precursor films, such as polyimides, resins, cellulose, and polyetherimides. Sazali et al. found that tubular carbon membranes with alumina powder as the intermediate layer showed excellent helium separation performance, with a He/N_2 selectivity of up to 472 [27].

However, physical adsorption of water and chemical adsorption of oxygen in the pores can significantly reduce membrane permeability. In practical applications, the long-term stability of carbon molecular sieve membranes against impurities in the feed gas must be considered to extend membrane lifespan and lower operating costs.

(3) Zeolite Molecular Sieve Membranes

Zeolite molecular sieve membranes feature uniform pore sizes comparable to molecular dimensions, excellent thermal stability, and chemical stability, allowing them to be used under high temperature and pressure. On the other hand, defects such as cracks, pinholes, and intergranular pores commonly arise during zeolite molecular sieve membrane synthesis, diminishing permeation separation performance and limiting their application in gas separation to some extent. Continuous progress in metal–organic framework materials and unique zeolite synthesis technologies is expanding the application potential of zeolite molecular sieve membranes [13].

China has recorded significant progress in the field of zeolite molecular sieve membranes. Led by Academician Xu Nanping, Nanjing Tech University has been researching zeolite molecular sieve membranes since the 1990s. They rolled out a technology for preparing hollow fiber all-silica zeolite molecular sieve membranes with independent intellectual property rights, whilst using domestic industrial-grade raw materials. These high-performance molecular sieve membranes are widely applied in water treatment fields such as ultrafiltration and reverse osmosis. The research team is currently exploring the application of zeolite molecular sieve membranes in helium separation from natural gas.

(4) Comparison of Inorganic Membrane Materials

Inorganic membrane materials exhibit good performance in helium separation. However, their structural fragility and complex production processes have limited their commercialization. Further efforts are needed to develop inorganic membranes with superior helium separation performance. The characteristics of different inorganic membrane materials are compared in Table 8.4.

Table 8.4 Comparison of different inorganic membranes

Membrane Types	Advantages	Disadvantages
Silicon dioxide membrane	Excellent chemical and thermal stability, usable at room temperature with low permeation flux	Usable only as tubular membranes, complex preparation process, high manufacturing cost
Carbon molecular sieve membrane	High gas permeability, good selectivity, strong chemical resistance, and thermal stability	Permeability significantly reduced by physical adsorption of water and chemical adsorption of oxygen
Zeolite molecular sieve membrane	Good chemical and thermal stability, usable under high temperature and pressure	Synthesis defects like cracks, pinholes, and intergranular pores which affect separation performance to a certain degree

8.3.3 Applications

With advancements in membrane preparation technologies and the development of novel synthesis and modification materials, both separation selectivity and permeability of membrane materials have significantly improved. Additionally, the cost of membrane separation components has decreased. Membrane technology has, therefore, become a technically and economically feasible option for separating helium from natural gas.

A typical membrane separation skid used in helium extraction plants is shown in Fig. 8.8.

8.4 Compressor

Compressors are devices that convert mechanical energy from a prime mover into gas pressure energy, which is used to increase gas pressure and transport gases. In the process for helium extraction from natural gas, multiple processes require a boost in gas pressure. The processes include permeate gas pressurization in membrane separation, refrigerant pressurization in cooling cycles, helium pressurization for storage and loading, and helium recovery.

The main types of compressors used include centrifugal compressors, reciprocating compressors, screw compressors, and diaphragm compressors, mostly used for pressurization and fluid transport. To be specific, centrifugal compressors are ideal for large flow rates and low pressure ratios, reciprocating compressors for

Fig. 8.8 Membrane separation skid

small flow rates and high pressure ratios, screw compressors for medium flow rates and pressure ratios, and diaphragm compressors for compressing gases with high helium concentrations.

8.4.1 Centrifugal Compressor

In helium extraction plants, centrifugal compressors are widely used for processes such as feed gas compression, mixed refrigerant compression, and export gas compression due to their ability to handle large gas volumes and high shaft power. The working principle of a centrifugal compressor involves gas entering the suction chamber, where the impeller rotates at high speed, performing work on the gas. The centrifugal force causes the gas to rotate, increasing its pressure, velocity, and temperature. The gas then enters the diffuser, where part of its kinetic energy is converted into pressure energy. A vacuum is created at the impeller, allowing continuous gas intake and discharge as the impeller rotates.

A centrifugal compressor mainly consists of a rotor and a stator. The structure of a single-stage centrifugal compressor is shown in Fig. 8.9.

Characteristics of centrifugal compressors [28]:

(1) Compact structure, small size, and light weight.
(2) Uniform and continuous discharge eliminates the need for intermediate tanks and similar devices.
(3) Low vibration, few wearing parts, no need for internal lubrication except for bearings, saving oil and avoiding compressed gas contamination.
(4) Easy speed regulation.
(5) High reliability, low maintenance requirements, and typically no need for backup units.
(6) Not resistant to impurities and liquid droplets.

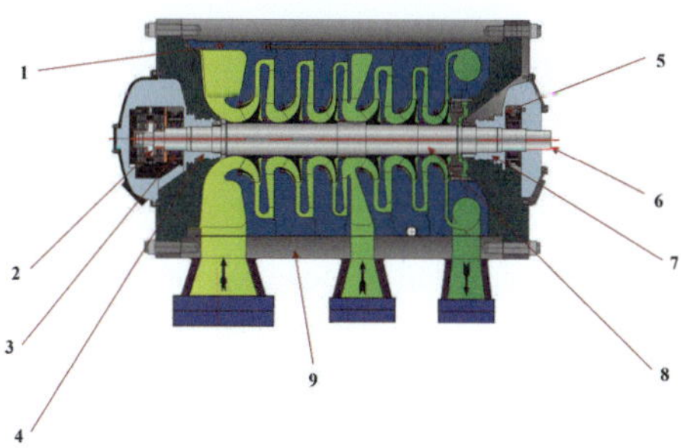

1 - Stator; 2 - Thrust Bearing; 3 - Radial Bearing; 4 - Dry Gas Seal; 5 - Radial Bearing; 6 - Coupling; 7 - Dry Gas Seal; 8 - Rotor; 9 - Cylinder

Fig. 8.9 Structure of single-stage centrifugal compressor

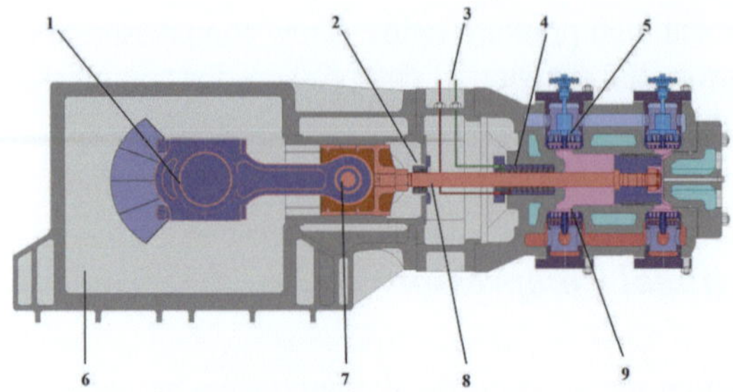

1 - Connecting Rod (connected to the crankshaft via the big end bearing); 2 - Oil Scraper Packing; 3 - Packing Injection; 4 - Piston Rod Packing; 5 - Intake Valve; 6 - Crankcase; 7 - Connecting Rod 8 - Piston Rod; 9 - Discharge Valve

Fig. 8.10 Structure of reciprocating compressor

8.4.2 Reciprocating Compressor

In helium extraction plants, reciprocating compressors are often used for small-volume, high-pressure applications, such as acid gas compression and nitrogen refrigeration. The working principle of a reciprocating compressor involves the linear reciprocating motion of a piston within a cylinder, changing the volume to pressurize and transport gases. The working cycle within the cylinder includes intake, compression, discharge, and expansion.

A reciprocating compressor mainly consists of a base and a working chamber, with its structure shown in Fig. 8.10. The base comprises the body, crankshaft, connecting rod, and crosshead assembly, among other components, and the working chamber includes the cylinder, piston, piston rod, piston ring, packing, air valve assembly, etc. [28].

Characteristics of reciprocating compressors:

(1) The gas handling capacity of a single unit is small, with a high discharge pressure.
(2) The operating speed is slow, only at medium and low speeds.
(3) Adaptable to different load conditions by changing the piston stroke.
(4) Stable discharge pressure, almost unaffected by gas flow regulation.
(5) Large in size and heavy in weight.
(6) Complex structure, regular maintenance required for wearing parts, significant maintenance workload, often requiring standby units.

8.4.3 Screw Compressor

In helium extraction plants, screw compressors are often used for low-flow BOG (Boil-Off Gas) compression and helium low-temperature refrigeration cycles. There are mature industrial products available with single-machine pressure ratios ranging from 4 to 15 and capacity displacement from 100 to 10,000 m^3/h [29]. The working process of a screw compressor includes three stages: suction, compression, and discharge.

Screw compressors are divided into single-screw and twin-screw ones. In a single-screw compressor, two star wheels that mesh with the screw are symmetrically arranged on both sides of the screw. The tooth slots, cylinder, and star wheels form elementary volumes that periodically expand and contract during the cyclic motion of the screw and star wheels, thereby achieving the suction, compression, and discharge processes. Twin-screw compressors consist of a pair of intermeshing male and female screws and a casing. During screw rotation, the lower part of the screws serves as the suction chamber, while the upper part acts as the compression and discharge chamber. Special air holes on the ends of the casing and end plates control the suction and discharge, maintaining a fixed internal volume ratio. Typically, when referring to screw compressors, it implies twin-screw compressors.

Based on their oil content during operation, screw compressors can be categorized into wet (oil-injected) screw compressors and dry (oil-free) screw compressors. In wet (oil-injected) screw compressors, lubricant oil or other liquids are injected into the working chamber to cool the compressed gas, lubricate the male and female rotors, enhance sealing performance, and achieve self-driving. In dry (oil-free) screw compressors, no liquid is injected into the working chamber, and the rotors maintain a certain gap through synchronous dimension adjustment without contact, ensuring that the compressed gas remains uncontaminated [28]. Typical structures are shown in Figs. 8.11 and 8.12.

Characteristics of screw compressors:

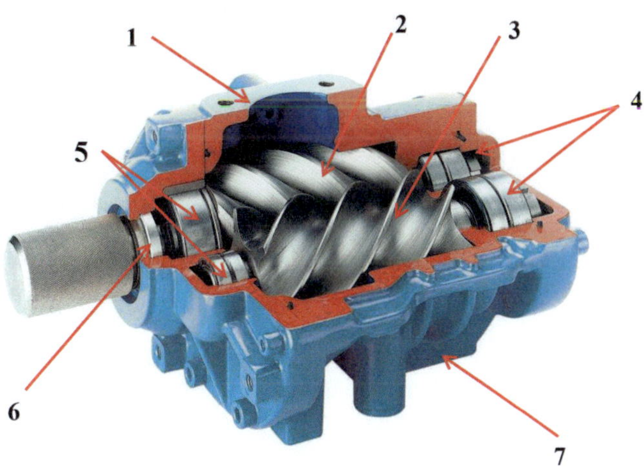

1 - Compressor Inlet; 2 - Male Rotor; 3 - Female Rotor; 4 - Radial and Thrust Bearings; 5 - Radial Bearing; 6 - Shaft Seal; 7 - Compressor Discharge Outlet

Fig. 8.11 Structure of wet (oil-injected) screw compressor [30]

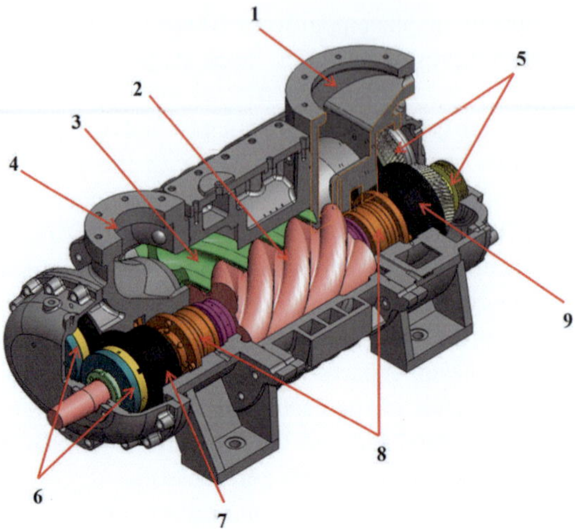

1 - Compressor Inlet; 2 - Male Rotor; 3 - Female Rotor; 4 - Compressor Discharge Outlet; 5 - Thrust Bearing; 6 - Radial Thrust Bearing; 7 - Seal; 8 - Radial Bearing; 9 - Synchronous Gear

Fig. 8.12 Structure of dry (oil-free) screw compressor

(1) Small handling capacity that typically ranges from 2 to 500 m^3/min.
(2) Outlet pressure: generally ≤ 1.5 MPa, with up to 4.5 MPa achievable in two-stage compression.
(3) Simple structure, small size, and light weight.
(4) Strong adaptability and certain tolerance to impurities, oil, and water in the gas.
(5) Few wearing parts, easy maintenance, no need for standby units.
(6) High speed with relatively high noise.

8.4.4 Diaphragm Compressor

Diaphragm compressors, also known as membrane compressors, offer the best sealing and lowest leakage rates among all types of compressors now available, making them ideal for compressing and transporting toxic, corrosive, flammable, explosive, radioactive, and precious gases [31]. In helium extraction plants, diaphragm compressors are used for pressurizing media with high helium concentration to prevent contamination and minimize leakage losses, such as in pure helium compressors, helium loading compressors, and helium recovery compressors.

The working principle of a diaphragm compressor involves compressing gas in a sealed diaphragm chamber, completely isolating the gas from lubricants and other external media through a diaphragm. The diaphragm itself performs reciprocating vibrations, and as it bends and deforms, the volume of the diaphragm chamber changes accordingly, functioning similarly to a compression piston. Due to the fact that the compression medium only comes into contact with the metal diaphragm, there is no need to introduce any lubricating substances into the cylinder. This eliminates the risk of oil contamination and ensures the purity of the compression medium. Additionally, due to the tight fastening of the diaphragm around its edges, the diaphragm chamber is completely sealed as a closed cavity, secured by several bolts around the cylinder head, hence its excellent sealing performance [28].

Figure 8.13 shows the membrane cavity of the diaphragm compressor, and Fig. 8.14 shows the compressor structure.

Characteristics of diaphragm compressors:

(1) High compression ratio, with pressure ranges extending above 25 MPa.

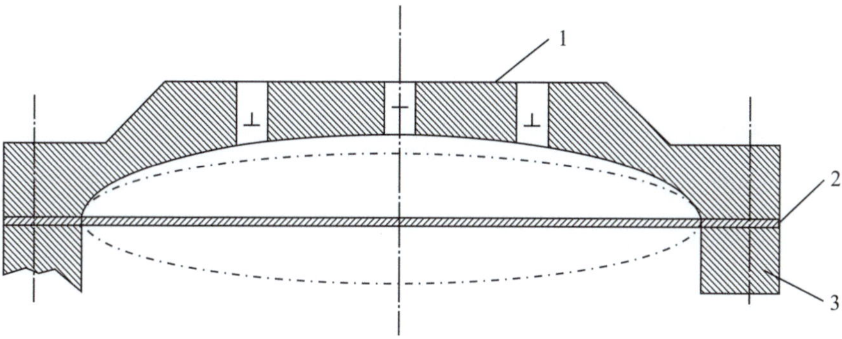

1 - Cover Plate; 2 - Diaphragm; 3 - Support Plate or Body

Fig. 8.13 Structure of diaphragm chamber

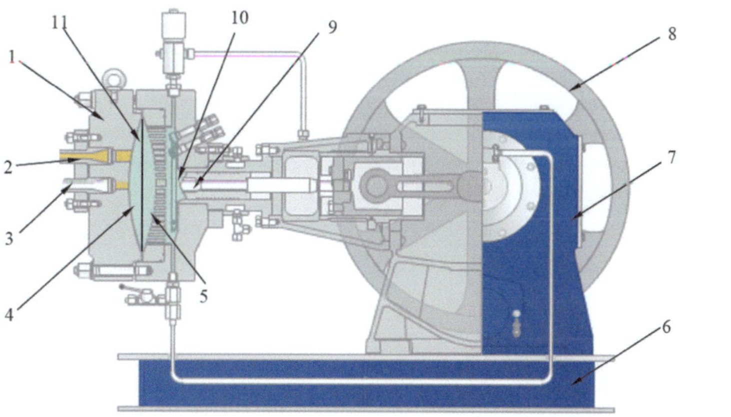

1 - Cylinder; 2 - Inlet; 3 - Outlet; 4 - Gas Chamber; 5 - Oil Chamber; 6 - Chassis; 7 - Crankcase; 8 - Pulley; 9 - Piston; 10 - Hydraulic Cylinder; 11 - Diaphragm

Fig. 8.14 Structure of diaphragm compressor

(2) Excellent sealing, with compressed gas free from contamination by lubricants or other solid impurities.
(3) Low noise, easy maintenance, and high reliability.
(4) Generally low gas handling capacity due to the limited deformation of the diaphragm.

8.4.5 *Applications*

In helium extraction projects, compressors are essential for pressurizing process gases at various locations. These applications include crude helium compression, gaseous helium filling, refrigeration cycles, and refrigerant pressurization. The selection of compressors should be based on the specific characteristics of the process and the pressurized medium. Table 8.5 presents the application of compressors in a typical million-cubic-meter-scale project for helium extraction from natural gas.

8.5 Helium Turboexpander

The primary function of an expander is to perform adiabatic expansion work, consuming the internal energy of the gas and significantly lowering its pressure and temperature for the purpose of refrigeration and cooling. Expanders are commonly used in cryogenic refrigeration and can be classified into piston expanders and turboexpanders based on their motion and structure. Adiabatic isentropic expansion is an important effect for obtaining low temperatures, and turboexpanders are the mechanical devices closest to achieving this process.

Turboexpanders can be classified into three categories based on their bearing lubrication methods: oil-lubricated bearing expanders, magnetic bearing expanders,

Table 8.5 Compressor applications in a typical million-cubic-meter-scale project for helium extraction from natural gas

Compressor name	Motor powerw/kW	Quantity/set	Type
Permeate gas compressor	1000	2	Reciprocating compressor
Mixed refrigerant compressor	5400	1	Centrifugal compressor
Nitrogen compressor	140	2	Reciprocating compressor
BOG compressor	160	1	Screw compressor
Carbon dioxide compressor	250	2	Reciprocating compressor
Crude helium compressor	55	2	Diaphragm compressor
Helium recovery compressor	30	2	Diaphragm compressor
Helium unloading compressor	40	2	Diaphragm compressor
Pure helium compressor	20	1	Diaphragm compressor

and gas bearing expanders. In the process for helium extraction from natural gas, turboexpanders are used for refrigeration during natural gas treatment and helium liquefaction. The expanders used for refrigeration in natural gas treatment generally employ oil-lubricated or magnetic bearings, which are mature and reliable products widely used in natural gas NGL recovery projects. The expanders used for refrigeration in helium liquefaction are gas bearing expanders. This section provides a brief introduction to the helium turboexpander.

8.5.1 Structure

A turboexpander utilizes the flowing velocity changes of the working medium to convert energy, thus it is also known as a velocity type expander. The working medium expands within the turboexpander to gain kinetic energy, which is then converted to external work, thereby reducing the internal energy and temperature of the medium at the expander outlet.

The main components of a turboexpander include volute, nozzle, guide vane, braking wheel, working wheel, diffuser, main shaft, and bearing seat. The most widely used type in cryogenic technology is the single-stage centripetal radial-axial flow reaction turboexpander with a semi-open working wheel, characterized by large enthalpy drops, high permissible speeds, simple structure, and high thermal efficiency. Its typical structure is shown in Fig. 8.15. When high-pressure gas passes through the volute, it is evenly distributed to the guide vanes, where the nozzle converts the thermodynamic energy into kinetic energy of the gas, reducing pressure, temperature, and enthalpy of the gas at the nozzle outlet [32].

Helium turboexpanders in cryogenic equipment typically operate with working medium temperatures typically below 40 K, with high expansion ratios. In low-flow

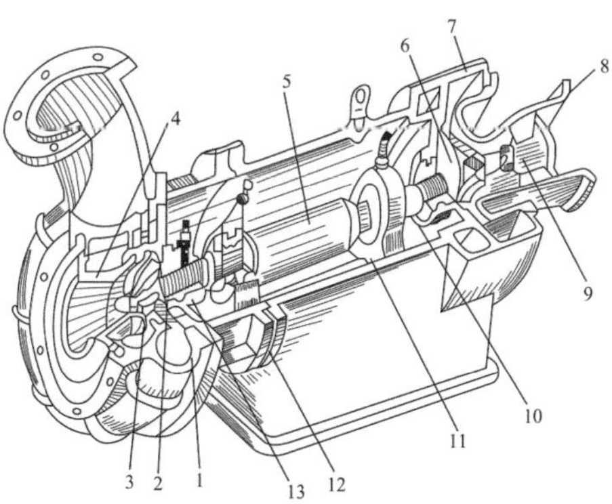

1 - Volute; 2 - Nozzle; 3 - Working Wheel; 4 - Diffuser;5 - Main Shaft; 6 - Fan Wheel; 7 - Fan Volute; 8 - Fan End Cover; 9 - Speedometer; 10 - Bearing Seat; 11 - Machine Body; 12 - Intermediate Body; 13 - Seal Sleeve

Fig. 8.15 Typical structure of single stage centripetal radial axial flow reaction turboexpander scenarios, the working wheel diameter is relatively small, the rotor of the expander can reach hundreds of thousands to nearly a million RPM, thus achieving high adiabatic efficiency. At such high speeds, traditional oil-lubricated bearings are inadequate due to inherent limitations and, on that account, must be replaced with gas bearings. Gas bearings are fluid film lubricated bearings that use gas as the lubricating medium. They offer a wide operating temperature range, eliminate contamination, and ensure smooth operation, among other advantages, and thus significantly enhance the reliability of turboexpanders, making them increasingly popular in cryogenic turboexpanders.

Given the special application scenarios, stable and efficient operation of helium turboexpanders is of particular importance. The turboexpanders are mainly supported by gas bearings, which may use the same gas medium as the cryogenic system to avoid cross-contamination. However, the low viscosity of gases results in lower damping and stiffness of gas bearings, making them prone to rotational instability at high speeds [33]. The cooling efficiency of a helium turboexpander largely depends on how closely the actual expansion index approaches the isentropic expansion process. The flow path within the helium turboexpander-including the inlet duct volute, expansion working wheel, nozzle, diffuser pipe, and braking fan wheel-can be validated and optimized using numerical simulation techniques.

8.5.2 Features

Helium turboexpanders with gas bearings have the following characteristics:

(1) High speed, low friction, and minimal friction heat, suitable for extremely high-speed operations.
(2) High precision with gas bearings that require only minimal clearances (generally below 50 um) to provide sufficient stiffness to support the main shaft.
(3) Suitable for light-load applications. Due to the low viscosity of gases, gas bearings have limited load-carrying capacity.

8.5.3 Applications

As widely applied key refrigeration equipment, helium turboexpanders have been studied and developed by companies such as Linde, Air Liquide, Hitachi Limited, and some Chinese institutions like the 101st Institute of the Sixth Academy of China Aerospace Science and Technology Corporation (CASC), and Technical Institute of Physics and Chemistry (TIPC) of the CAS. These developments cater to different functional needs, including large variable capacity helium turboexpanders, high-expansion-ratio helium turboexpanders, gas–liquid phase helium turboexpanders, and micro-refrigerator helium turboexpanders, among others. The Linde helium turboexpander is shown in Fig. 8.16.

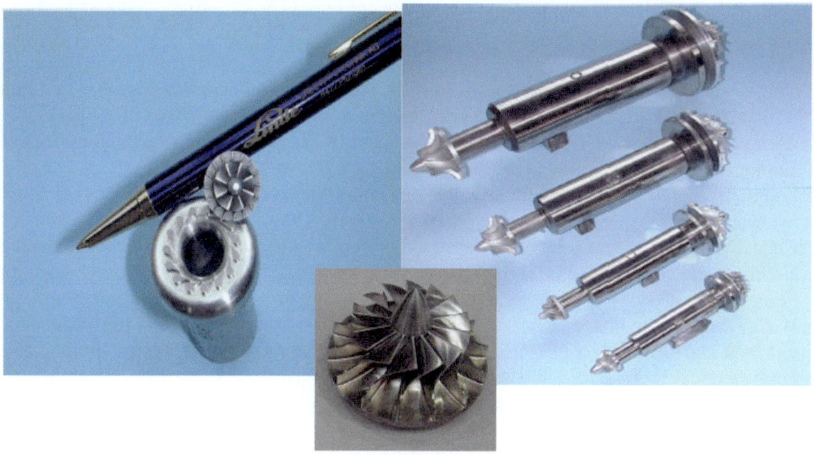

Fig. 8.16 Linde helium turboexpander.

Fig. 8.17 Helium turboexpander from the TIPC of the CAS [34]

For years, the TIPC of the CAS has been dedicated to the research of key technologies in large-scale cryogenic systems. They have addressed core technical challenges in areas such as long-piston expanders and gas bearing turboexpanders and have developed and applied a turbo-expansion refrigeration helium liquefier with a capacity of 300 L/h. The helium turboexpander developed by the TIPC of the CAS is shown in Fig. 8.17.

References

1. Chinese Association of Refrigeration. Research report on strategic development of China's refrigeration industry. Beijing: China Architecture & Building Press; 2016 (in Chinese).
2. Zheng X, Wang GH, Chen J, et al. Development of vacuum brazing technology for 6061 aluminum alloy. Light Alloy Fabricat Technol. 2014;2(1):8–12 (in Chinese).
3. Zohuri B, Zohuri B. Cryogenic technologies. In: Hybrid energy systems: driving reliable renewable sources of energy storage; 2018. pp. 39–88.
4. Chen ZY, Zhao L, Zhu QJ, et al. Comparison of domestic and foreign technical standards for aluminum plate-fin heat exchangers and introduction to NB/T 47006-2019 revisions. Petrochem Equip. 2020;49(3):45–51 (in Chinese).
5. Wang HS. Plate-fin heat exchanger. Beijing: Chemical Industry Press; 1984 (in Chinese).
6. Lanzhou Petroleum Machinery Research Institute. Modern tower technology. Beijing: China Petrochemical Press; 2005 (in Chinese).
7. Li SY, Wang WJ. Research progress on tray technology application. Technol Dev Chem Ind. 2011;40(12):30–4 (in Chinese).
8. Chen DC, Liu NH. New advances in modern packed column separation technology. Shanghai Chem Ind. 1994;19:22–25 (in Chinese).
9. Ruban GN, Martynov VG, Mkrtychan YS. Helium. Reserves, extraction, storage, transport and application. ИЗДАТЕЛЬСКИЙ ЦЕНТР РГУ нефти и гаэа (НИУ) имени И.М. Губкина; 2018.
10. Scholes CA, Stevens GW, Kentish SE. Membrane gas separation applications in natural gas processing. Fuel. 2012;96:15–28.
11. Weller S, Steiner WA. Separation of gases by fractional permeation through membranes. J Appl Phys. 1950;21(4):279–283.
12. Stern SA, Sinclair TF, Gareis PJ, et al. Helium recovery by permeation. Ind Eng Chem. 1965;57(2):49–60.
13. Lu YB. Research progress of extracting helium from natural gas with membrane separation. Petrochem Technol. 2020;49(5):513–518 (in Chinese).
14. Adewole JK, Sultan AS. Polymeric membranes for natural gas processing: polymer synthesis and membrane gas transport properties, vol. 1. Cham: Springer; 2019. pp. 1–37
15. Adewole JK, Sultan AS. Polymeric membranes for natural gas processing: polymer synthesis and membrane gas transport properties. Functional polymers. Cham: Springer, 2019: 941–976.
16. Park J, Lee KS. A two-dimensional model for the spiral wound reverse osmosis membrane module. Desalination. 2017;416:157–165.
17. Yang MZ, Li L, Wu HY, et al. Research progress on performance analysis and optimization of spiral-wound membrane modules. Membr Sci Technol. 2022;3:187–194 (in Chinese).
18. Chen Y, Wang CH, Wu M. Gas membrane separation technology and applications. Beijing: Chemical Industry Press; 2004 (in Chinese).
19. Zheng PJ, Xie W, Bai J, et al. Research progress on gas separation membrane technology for helium extraction from natural gas. Membr Sci Technol. 2022;6:168–177 (in Chinese).
20. Vaezi MJ, Bayat Y, Babaluo AA, et al. Separation of helium from gases using the synthesized hydroxyl sodalite membrane. Scientia Iranica. 2016;23(3):1136–1143.
21. Gantzel PK, Merten U. Gas separations with high-flux cellulose acetate membranes. Ind Eng Chem Process Des Dev. 1970;9(2):331–332.
22. Kumazawa H, Sada E, Nakata K, et al. Enrichment of helium by asymmetric hollow-fiber membrane of cellulose triacetate. J Appl Polym Sci. 1994;53(1):113–119.
23. López-González M, Saiz E, Riande E, et al. Transport of helium in polycarbonate membranes. Polymer. 2002;43(2):409–413.
24. Min KE, Paul DR. Effect of tacticity on permeation properties of poly (methyl methacrylate). J Polym Sci, Part B: Polym Phys. 1988;26(5):1021–1033.
25. Choi JI, Jung CH, Han SH, et al. Thermally rearranged (TR) poly (benzoxazole-co-pyrrolone) membranes tuned for high gas permeability and selectivity. J Membr Sci. 2010;349(1–2):358–368.

26. Shelekhin AB, Dixon AG, Ma YH. Adsorption, permeation, and diffusion of gases in microporous membranes. II. Permeation of gases in microporous glass membranes. J Membr Sci. 1992;75(3):233–244.
27. Sazali N, Salleh WNW, Ismail AF, et al. Influence of intermediate layers in tubular carbon membrane for gas separation performance. Int J Hydrogen Energy. 2019;44(37):20914–20923.
28. Yu YZ, Jiang PZ, Sun SY. Compressor engineering handbook. Beijing: China Petrochemical Press; 2012 (in Chinese).
29. Hu ZJ, Wu XJ, Lin WJ, et al. Research, development, and application of core technology for large helium screw compressors. Chin Sci Bull. 2022;67(21):2482–2491 (in Chinese).
30. Larralde E, Ocampo R. Selection of gas compressors: part 2. World Pumps. 2011;2011(9):36–43.
31. Wang H. Discussion on diaphragm compressor issues. Equip Manag Maint. 2023;6:82–83 (in Chinese).
32. Chen GB. New cryogenic technology. Shanghai: Shanghai Jiao Tong University Press; 2003 (in Chinese).
33. Hou Y, Wang J, Xiong LY, et al. Application of gas bearings in cryogenic turboexpanders. Cryogenics. 2001;3:29–33 (in Chinese).
34. Liu LQ. Progress in the development of large helium cryocoolers. Vac Cryog. 2020;6:471–475 (in Chinese).

Chapter 9
Helium Extraction Plants from Natural Gas

Haiyang Long and Jianhua Tan

The helium extraction industry from natural gas has a history spanning over a century, with large-scale helium extraction plants primarily located in the United States, Qatar, Russia, and Australia, among other countries. This chapter introduces the construction scale, technical characteristics, and main processes of typical helium extraction plants, based on the background and resource conditions of different countries.

9.1 Typical Helium Extraction Plants out of China

Countries with abundant helium resources and high concentrations typically use cryogenic technologies for helium extraction or extract helium from the BOG of large LNG plants. These helium extraction facilities are large-scale and boast high helium production capacities. This section briefly introduces typical helium extraction plants located in Russia, Qatar, Australia, Canada, and Poland.

9.1.1 Russia's Amur Gas Processing Plant

The helium extraction facility at Russia's Amur Gas Processing Plant (AGPP) is an integral part of AGPP. The gas source for AGPP comes from the Yakutia and Irkutsk central gas processing stations, transported via the "Power of Siberia" pipeline to

H. Long (✉)
China Petroleum Engineering & Construction Corporation Southwest Company, Chengdu, China
e-mail: longhaiy_sw@cnpc.com.cn

J. Tan
PetroChina Tarim Oilfield Company, Korla, Xinjiang, China
e-mail: tanjh-tlm@petrochina.com.cn

L. Pu (ed.), *Helium Engineering Technologies*,
https://doi.org/10.1007/978-981-96-4086-7_9

Fig. 9.1 AGPP processing plant facilities

Blagoveshchensk, the capital of Amur Oblast in Russia's Far East, near the border with China, for processing and helium extraction.

AGPP is currently the largest gas processing plant in Russia and is the heart of the China-Russia eastern energy transmission system. It has a feed gas processing capacity of 4.2×10^{10} m^3/a, a commercial natural gas production capacity of 3.8×10^{10} m^3/a, and a helium production capacity of 6×10^7 m^3/a. The AGPP processing plant is shown in Fig. 9.1.

1. Overall Process Flow

AGPP is equipped with six natural gas processing units, including three helium purification units. The helium extraction technology and core equipment are provided by Linde. The helium content in the helium-containing natural gas at AGPP is about 4000 mL/m^3 (ppmV). The plant uses the "ethane recovery combined with helium extraction" cryogenic technology route. The gas separated at the top of the demethanizer in the ethane recovery unit sequentially enters the first and second denitrification columns for nitrogen removal and helium concentration. The main process units of the helium extraction system include ethane recovery, crude helium extraction, crude helium refining, and helium liquefaction and storage facilities. The overall process flow of the AGPP plant is shown in Fig. 9.2.

2. Main Process Units

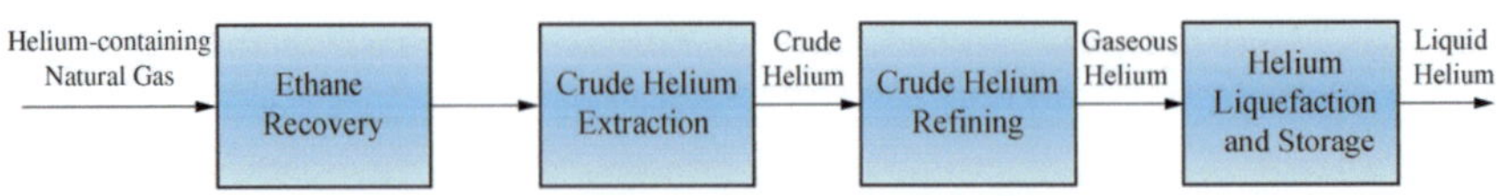

Fig. 9.2 Process flow diagram of AGPP processing plant

(1) Ethane Recovery Unit

The pre-treated purified natural gas, after pre-cooling and separation, enters the demethanizer for rectification separation. The top gas of the demethanizer (methane, ethane, helium, nitrogen, and hydrogen) enters the downstream crude helium extraction unit for further helium extraction. The liquid phase at the bottom of the demethanizer enters the deethanizer for rectification production of ethane. The liquid phase at the bottom of the deethanizer enters the fractionation unit to separate propane, butane, and condensate products.

(2) Crude Helium Extraction Unit

The gas atop the demethanizer enters the first denitrification column for primary helium concentration. The liquid phase at the bottom of the first denitrification column (containing methane, ethane, and nitrogen) throttles to provide cooling for the cold box. The gas phase atop the first denitrification column (containing helium, hydrogen, nitrogen, and methane) condenses and then enters the second denitrification column for further helium concentration. The liquid phase at the bottom of the second denitrification column (containing nitrogen and methane) gets pressurized and recycled as reflux for the top of the first denitrification column. The gas phase atop the second denitrification column is crude helium, reheated to ambient temperature before entering the crude helium refining unit.

(3) Crude Helium Refining Unit

The concentrated crude helium sequentially undergoes catalytic combustion and PSA to remove hydrogen, nitrogen, oxygen, and other impurities, yielding high-purity helium.

(4) Helium Liquefaction and Storage

High-purity helium is condensed into liquid helium products through a helium liquefier, stored in liquid helium tank containers.

9.1.2 *Qatar's Helium Extraction Plants*

Qatar holds the second-largest helium reserves after the United States. The helium content in natural gas from Qatar's North Field is approximately 490 mL/m^3 (ppmV). Typically, recovering helium directly from natural gas at such low concentrations is uneconomical. Nevertheless, during the natural gas liquefaction process, helium is enriched in the BOG to about 2%, thereby increasing the economic viability of helium recovery [1].

Currently, Qatar has built three large helium extraction plants [2], as detailed in Table 9.1.

Table 9.1 Basic information on Qatar's helium extraction plants

Plant name	Product	Liquid helium production/(10^7 m^3/a)	Helium extraction process	Commissioning year	Helium source
Qatar helium extraction plant (phase 1)	Liquid helium	1.864	Helium extraction from LNG denitrification BOG	2005	QatarGas LNG1-3 RasGas LNG1-4
Qatar helium extraction plant (phase 2)	Liquid helium	3.755	Helium extraction from LNG denitrification BOG	2013	QatarGas LNG4-7 RasGas LNG6-7
Qatar helium extraction plant (phase 3)	Liquid helium	1.2	Helium extraction from GTL denitrification BOG	2021	–

Qatar currently produces about one-third of the world's helium supply. With the gradual commissioning of the North Field East (NFE) and North Field South (NFS) expansion projects, Qatar's LNG production capacity is expected to reach 1.26×10^8 t/a by 2027, further increasing helium production capacity.

1. Overall Process Flow

Qatar uses denitrification BOG from LNG or Gas-to-Liquids (GTL) plants as the helium source, following the principle of decentralized extraction and centralized refining. The following is a brief introduction to Qatar Helium Extraction Plant (Phase 2) [3].

The goal of Qatar Helium Extraction Plant (Phase 2) is to extract helium from RasGas LNG production lines 6–7 and Qatargas LNG production lines 4–7 and produce liquid helium for sale. Six extra-large LNG production lines share one set of helium extraction facilities. The helium extraction process is self-refrigerated, relying on the Joule–Thomson (J–T) effect in helium recovery heat exchangers and flash tanks. The helium extraction technology is provided by Air Products and Chemicals Inc. The main process units of the helium extraction system include cryogenic distillation, catalytic dehydrogenation, PSA, and helium liquefaction and storage facilities. The process flow of Qatar Helium Extraction Plant (Phase 2) is shown in Fig. 9.3.

2. Main Process Units

(1) Cryogenic Distillation Unit

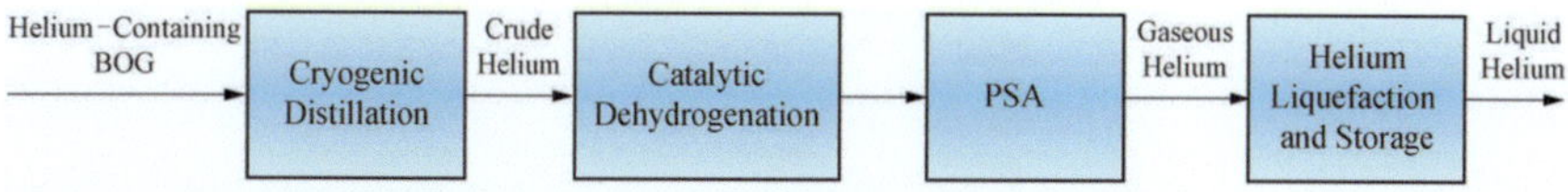

Fig. 9.3 Process flow diagram of Qatar helium extraction plant (phase 2)

The upstream BOG, with a helium content of about 2%, enters the cryogenic distillation unit. The BOG is cooled through a cold box, forming a two-phase flow that enters the helium recovery flash tank for multistage gas–liquid separation, resulting in crude helium with about 50% helium content. The crude helium then enters the downstream catalytic dehydrogenation unit. The condensate flowing out from the bottom of the helium recovery flash tank mainly consists of methane and nitrogen, which, after cooling recovery, is sent to fuel gas.

(2) Catalytic Dehydrogenation Unit

The crude helium from the cryogenic distillation unit enters the catalytic dehydrogenation unit. A certain amount of air is added to the crude helium feed gas, and under the action of a catalyst, the hydrogen in the crude helium reacts with the oxygen in the air and is removed.

(3) PSA Unit

The crude helium, after catalytic dehydrogenation, enters the PSA unit, where methane, oxygen, nitrogen, and other impurities in the crude helium are adsorbed by the adsorbent in the PSA towers, resulting in gaseous helium products. Part of the gaseous helium is loaded for transport, while another part is directed to the helium liquefaction unit to produce liquid helium.

(4) Helium Liquefaction and Storage Unit

The gaseous helium from the PSA unit enters the helium liquefier, where it is deeply cooled to produce liquid helium.

The plant is equipped with four liquid helium tank containers, as shown in Fig. 9.4, and each loading platform is equipped with a liquid helium tank container with a design capacity of approximately 120 m^3. Additionally, the plant has truck loading and weighing facilities for safe and efficient liquid helium operations. The plant also features a liquid helium BOG recovery system, including eight gas holders, each with a volume of 200 m^3. The helium recovery compressors are reciprocating compressors, capable of collecting, storing, and recompressing all continuous and intermittent helium emissions.

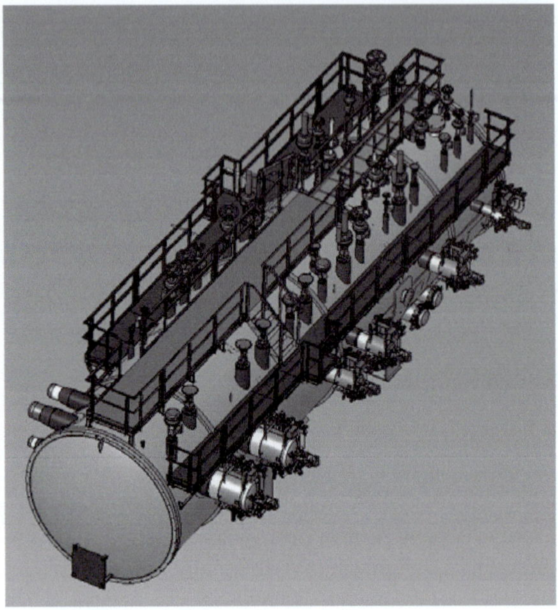

Fig. 9.4 Helium Liquefier [4]

9.1.3 Darwin Helium Extraction Plant in Australia

The Darwin Helium Extraction Plant in Australia was completed and commissioned in 2010. Located on Palmerston Island near Darwin in northern Australia, it has a liquid helium production capacity of approximately 4.3×10^6 m^3/a, making it the first liquid helium production site in the Southern Hemisphere. About one-third of the liquid helium produced by the plant is used in the Australian market, with the remainder mainly exported to the Asia–Pacific region [5].

1. Overall Process Flow

The helium content in natural gas from the Bonaparte Basin is approximately 400 mL/m^3 (ppmV). After natural gas liquefaction and concentration, the BOG has a helium concentration of 3%, which is then sent to the Darwin Helium Extraction Plant.

The Darwin Helium Extraction Plant employs a process that includes cryogenic rectification, catalytic dehydrogenation, cryogenic condensation, PSA, helium liquefaction, and storage. The process flow is shown in Fig. 9.5.

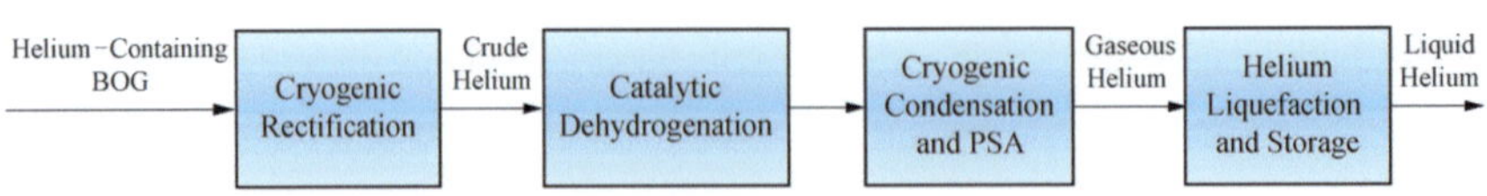

Fig. 9.5 Process flow diagram of darwin helium extraction plant

2. Main Process Units

(1) Cryogenic Rectification Unit

The low-pressure helium-containing BOG is first pressurized by a first-stage feed gas compressor to 0.2 MPa, then enters a first-stage denitrification heat exchanger to cool to about − 193 °C for partial condensation. In the separator, gas–liquid separation occurs; due to the low boiling point of helium, almost all helium remains in the gas phase, increasing its concentration to 26%. The gas is reheated to ambient temperature and pressurized to 3.1 MPa by a second-stage feed gas compressor. The coalescer and oil absorber remove any lubricating oil carried in the crude helium, and the purified gas is sent to the downstream catalytic dehydrogenation unit.

(2) Catalytic Dehydrogenation Unit

The crude helium from the cryogenic rectification unit enters the catalytic dehydrogenation unit. Under the action of a catalyst, hydrogen is removed through an oxidation reaction.

(3) Cryogenic Condensation and PSA Unit

The dehydrogenated crude helium enters the cryogenic condensation and PSA unit. After cooling to − 193 °C in a primary heat exchanger, it enters a primary separator where the helium concentration in the gas phase coming out of the top climbs to 93%. It then enters a secondary heat exchanger, cools to − 205 °C, and undergoes gas–liquid separation in a secondary separator to yield 99% crude helium. This crude helium is reheated to ambient temperature in the secondary heat exchanger before being sent to the PSA tower, producing high-purity helium with a purity of 99.999%.

(4) Helium Liquefaction and Storage Unit

The helium liquefier, provided by Linde, uses an improved Brayton cycle process with liquid nitrogen precooling. Liquid helium is stored in helium tank container with a capacity of approximately 120 m^3, and the plant is equipped with two liquid helium loading systems.

9.1.4 *Mankota Helium Extraction Plant in Canada*

The Mankota gas field is a large anticline gas field located in southwestern Saskatchewan, Canada. The Mankota Helium Extraction Plant, completed in August 2016, utilizes Linde's helium extraction technology and is the first plant in Canada to use membrane separation and PSA units to produce 99.999% helium. The project uses SEPURAN polymer membranes from Germany's Evonik Industries, which effectively separate methane, nitrogen, helium, and hydrogen from mixed gases.

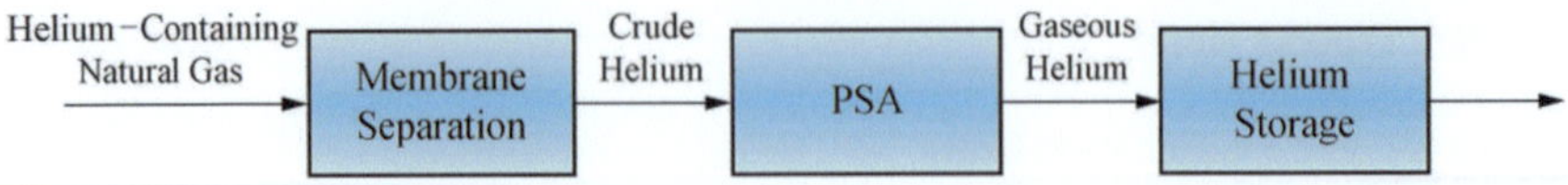

Fig. 9.6 Process flow diagram of mankota helium extraction plant

The raw natural gas is designed with a processing capacity of 2.8×10^5 m^3/d, with a helium content of 1%–2%. The plant employs an ambient temperature process of "membrane separation + PSA" for helium extraction. Helium-containing natural gas enters the membrane separation unit, where the permeate gas is concentrated and pressurized before entering the downstream PSA unit to remove methane, oxygen, nitrogen, and other impurities, resulting in 99.999% high-purity helium. The helium product is then loaded for transport. The process flow is shown in Fig. 9.6.

9.1.5 Odolanów Helium Extraction Plant in Poland

The Odolanów Helium Extraction Plant in Poland started operations in 1977, processing helium-containing natural gas at a rate of 3.5×10^6 m^3/d, with a helium content of approximately 0.4% and a nitrogen content as high as 42.75% [6, 7].

1. Overall Process Flow

The plant uses "cryogenic rectification" technology to extract helium and produce liquid helium. The main processes at the Odolanów Helium Extraction Plant include helium-containing natural gas pretreatment, cryogenic rectification, catalytic dehydrogenation, cryogenic condensation, and helium liquefaction and storage. The process flow is illustrated in Fig. 9.7.

2. Main Process Units

(1) Helium-Containing Natural Gas Pretreatment Unit

The pretreatment of helium-containing natural gas involves amine scrubbing for desulfurization and decarbonization, molecular sieve dehydration, and adsorption for heavy hydrocarbon removal.

(2) Cryogenic Rectification Unit

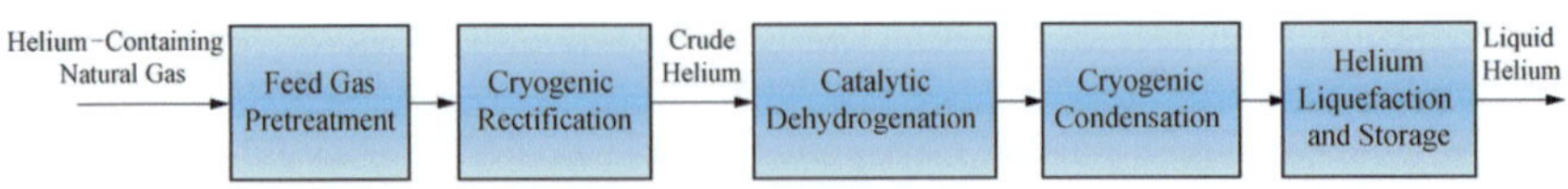

Fig. 9.7 Process flow diagram of Odolanów helium extraction plant

The cryogenic rectification unit employs the classic double-column process widely used in the air separation industry. This low-temperature process does not require any external cooling sources, hence its low operational costs.

The pretreated natural gas is pre-cooled in the cold box and then throttled to 2.7 MPa before entering the lower section of the high-pressure rectification column (lower column) for initial separation of methane and nitrogen components, with helium mainly concentrating in the nitrogen-rich gas at the top of the column for preliminary concentration. The methane-rich liquid from the bottom of the high-pressure rectification column (lower column) is throttled and enters the middle of the low-pressure rectification column (upper column). The condenser of the high-pressure rectification column (upper column) and the reboiler of the low-pressure rectification column (lower column) are integrated. The nitrogen-helium mixture from the top of the high-pressure rectification column (lower column) undergoes partial condensation in the condenser atop the column, followed by gas–liquid separation. The liquid phase returns to the top of the high-pressure rectification column (lower column) as reflux, while the gas phase, with helium concentration around 10%, undergoes further condensation and enters the low-pressure rectification column (lower column). The cooling required for nitrogen liquefaction in the condenser of the high-pressure rectification column (lower column) is mainly provided by the evaporation of liquid methane at the bottom of the low-pressure rectification column (upper column). Two liquid streams are drawn from the middle of the high-pressure rectification column (lower column) to serve as feed and top reflux for the low-pressure rectification column (upper column). The liquid methane at the bottom of the low-pressure rectification column (upper column), with a nitrogen content of about 4%, is boosted by a pump, exchanged with the feed gas, vaporized, and then sent to the export pipeline. The gas phase at the top of the low-pressure rectification column (upper column), with crude helium concentration increased to about 88%, is subject to heat recovery and then exported.

(3) Catalytic Dehydrogenation Unit

Crude helium enters the catalytic dehydrogenation unit, where hydrogen is removed through a reaction with oxygen under the action of a platinum catalyst. The water produced by the reaction is removed using molecular sieve adsorption.

(4) Cryogenic Adsorption Unit

In the cryogenic condensation unit, crude helium is condensed under vacuum using liquid nitrogen and purified by activated carbon adsorption to remove nitrogen, oxygen, and other impurities, yielding high-purity helium.

(5) Helium Liquefaction and Storage Unit

The high-purity helium is liquefied and stored in liquid helium tank containers, with the capacity to store approximately seven days' worth of production.

9.2 Typical Helium Extraction Plants in China

China's helium extraction industry began in the 1960s. With the development of mature technologies for crude helium extraction from natural gas, crude helium refining, and gaseous helium storage and transportation, these technologies basically meet the needs of large-scale construction of helium extraction projects. According to statistics, there are currently more than 20 enterprises in China with the capability to extract helium. As of 2023, the total production capacity has exceeded 6×10^6 m^3, though the actual output was about 3×10^6 m^3.

This section offers brief introductions to three typical helium extraction plants: Rongxian Helium Extraction Plant of Chengdu Natural Gas Chemical Complex, BOG Helium Extraction Plant of Inner Mongolia Xingsheng Natural Gas Co., Ltd., and BOG Helium Extraction Plant of Yulin Yuanheng Energy Co., Ltd.

9.2.1 Rongxian Helium Extraction Plant of Chengdu Natural Gas Chemical Complex

In the 1960s, China established its first helium extraction test unit—Weiyuan Helium Extraction Test Unit I in Weiyuan, with a designed capacity of 5×10^4 m^3/d and an annual helium production of 2×10^4 m^3. In 1989, the Weiyuan Helium Extraction Test Unit II was built with a designed capacity of 1×10^5 m^3/d and an annual helium production of 4×10^4 m^3 [8].

In 2012, the Rongxian Helium Extraction Plant of Chengdu Natural Gas Chemical Complex was established in Dongxing Town, Sichuan Province, as shown in Fig. 9.8. The plant has a designed capacity of 4×10^5 m^3/d, with a helium content in natural gas of about 1800 mL/m^3 (ppmV). The plant's helium production capacity is about 2.1×10^5 m^3/a, with crude helium purity ranging from 90% to 95%. Crude helium is transported by cylinder trailers to Chengdu Natural Gas Chemical Complex for further purification and refining, yielding helium with a purity of 99.999% [8].

1. Overall Process Flow

The helium-containing natural gas at Rongxian Helium Extraction Plant undergoes a decarbonization and dehydration pretreatment system. After decarbonization, the natural gas undergoes dehydration to prevent freezing blockages caused by carbon dioxide and water in the downstream cryogenic section. The pretreated helium-bearing natural gas then enters the demethanization unit, which primarily uses the "expansion + nitrogen cycle refrigeration process". The front-end refrigeration uses an expander, while the deep low-temperature concentration section uses nitrogen cycle refrigeration technology, achieving a temperature as low as − 190 °C. The crude helium from the demethanization unit enters the catalytic dehydrogenation unit, where a certain amount of air is mixed with the crude helium feed gas. Under

Fig. 9.8 Rongxian helium extraction plant

the action of a palladium alloy catalyst, the oxygen in the air reacts fully with the hydrogen in the crude helium. After that, the crude helium, with a purity of 90%–95%, is transported by cylinder trailers to Chengdu Natural Gas Chemical Complex, where it enters the helium purification unit. Following high-pressure low-temperature adsorption, it yields a high-purity product with a purity of 99.999%. The overall process flow is shown in Fig. 9.9.

2. Main Process Units

(1) Decarbonization and Dehydration Unit

Helium-bearing natural gas first enters the decarbonization tower, where MDEA solution is used for decarbonization, reducing carbon dioxide to 50 mL/m^3 (ppmV) or lower. The MDEA-rich solution absorbing carbon dioxide then enters the regeneration tower for regeneration. The decarbonized natural gas enters the molecular

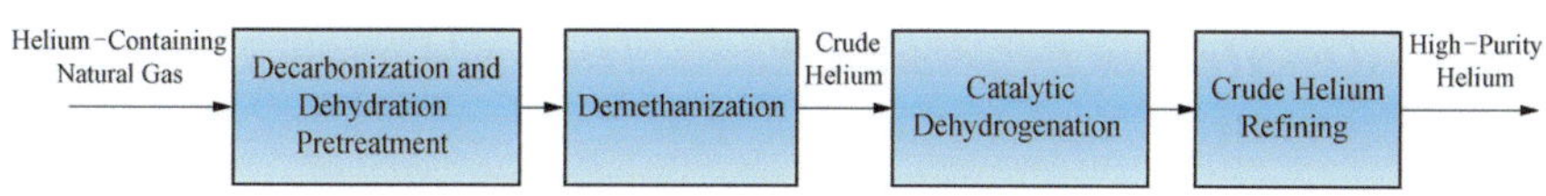

Fig. 9.9 Process flow diagram of rongxian helium extraction plant

sieve dehydration system for dehydration. A three-tower process is used for dehydration. In comparison to the traditional two-tower molecular sieve dehydration process, the heating furnace runs continuously, and the high-temperature natural gas directly enters the heating furnace after cold blow, thereby lowering regeneration energy consumption.

(2) Demethanization Unit

The pretreated helium-containing natural gas enters the demethanization unit's main cold box. After cold exchange, it enters the first-stage concentration column. The bottom stream of the first-stage concentration column mainly returns to the main cold box for cooling recovery. Subsequently, it expands and re-enters the main cold box to recover cooling again, and then forms medium-pressure tail gas, which is pressurized by a coaxial booster and sent to the natural gas export pipeline. The helium-bearing natural gas from the top of the first-stage concentration column enters the second-stage concentration column for separation. At the top of the second-stage concentration column, liquid nitrogen is used as the separation cooling source. After separation, the crude helium obtained has a concentration of about 75%.

(3) Catalytic Dehydrogenation Unit

The crude helium from the demethanization unit enters the catalytic dehydrogenation unit. A small amount of air is mixed into the crude helium feed gas through a trace air mixer. Under the action of a palladium catalyst, hydrogen is removed from the crude helium through a reaction in the first-stage reactor. Then the gas enters the first-stage cooler, where it is cooled to 40 °C by heat exchange with cooling water. The water produced by the reaction is cooled and then separated in a separator. After going through four rounds of air mixing, reaction, cooling, and water separation, the hydrogen content in the crude helium is reduced to less than 1 mL/m^3 (ppmV), yielding helium with a purity of 90%–95%.

(4) Cryogenic Adsorption Unit

The crude helium obtained at Rongxian Helium Extraction Plant is transported by cylinder trailers to Chengdu Natural Gas Chemical Complex, where it enters the helium refining unit. The crude helium is cooled to approximately − 180 °C using liquid nitrogen for low-temperature adsorption, which removes oxygen, neon, argon, and other impurity gases. A high-purity product with a purity of 99.999% is achieved.

9.2.2 BOG Helium Extraction Plant of Inner Mongolia Xingsheng Natural Gas Co., Ltd.

The BOG Helium Extraction Plant of Inner Mongolia Xingsheng Natural Gas Co., Ltd. is located in the Erdos New Energy Industry Demonstration Park, Xini Town,

Fig. 9.10 BOG helium extraction plant of Inner Mongolia Xingsheng Natural Gas Co., Ltd.

Hangjin Banner, east to the construction land of Hangjin Banner Hengdong Natural Gas Co., Ltd., as shown in Fig. 9.10. The BOG raw material comes from the LNG plant of Xingsheng Natural Gas Co., Ltd., which has a natural gas processing capacity of 2.6×10^6 m^3/d with a helium content of 1100–1200 mL/m^3 (ppmV). The BOG produced by the LNG plant has a helium content of 1%–2%. The BOG Helium Extraction Plant produces both gaseous and liquid helium products. It was commissioned in August 2020 and has a helium product output of about 2500 m^3/d.

1. Overall Process Flow

The BOG crude helium extraction unit uses low-temperature separation to extract helium from the BOG and separates liquid nitrogen as the refrigeration source for the helium refining unit. The crude helium is temporarily stored in a crude helium holder and then enters the catalytic dehydrogenation unit. Under the action of a palladium alloy catalyst, hydrogen reacts fully with added excess oxygen to form water, which is then removed by molecular sieve dehydration towers. The crude helium further enters the helium refining unit, where it is pressurized and sent to the low-temperature adsorption cold box. Under the cooling effect of liquid nitrogen (obtained from the crude helium extraction unit), activated carbon adsorbs nitrogen, oxygen, and other impurities from the crude helium to obtain 99.999% gaseous helium, which is then stored in a gaseous helium storage tank. The gaseous helium can enter the helium liquefaction unit for further liquefaction, resulting in liquid helium. The overall process flow is shown in Fig. 9.11.

2. Main Process Units

(1) Crude Helium Extraction Unit

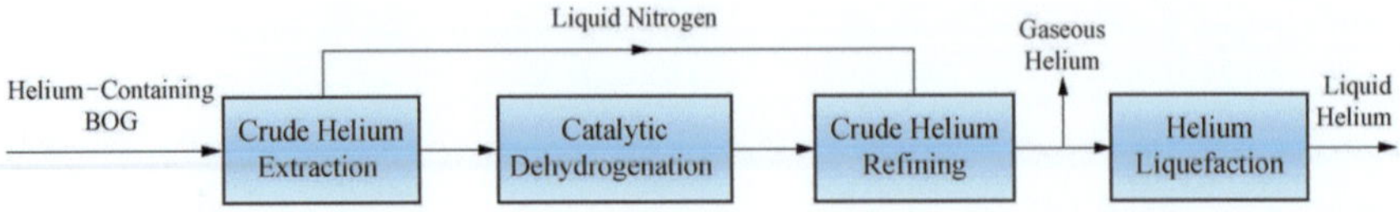

Fig. 9.11 Process flow diagram of BOG helium extraction project, Inner Mongolia Xingsheng Natural Gas Co., Ltd.

The BOG is pressurized to 5 MPa using the BOG booster of the original LNG plant. Then, it enters the crude helium extraction cold box, which contains two sets of rectification columns. The primary rectification separates nitrogen and helium to obtain crude helium (about 80%), and the secondary rectification separates nitrogen and methane from the bottom product of the primary rectification column, obtaining liquid nitrogen. The liquid nitrogen is stored in a liquid nitrogen storage tank and provides the necessary liquid nitrogen for the helium refining unit. The cooling in this unit mainly comes from the mixed refrigerant and nitrogen refrigerant of the LNG production unit. The crude helium extraction unit is equipped with a refrigeration expander.

(2) Catalytic Dehydrogenation Unit

The crude helium is first preheated in a preheater and then mixed with added oxygen before entering the first-stage dehydrogenation reactor, with an outlet temperature of around 200 °C. Afterward, the gas undergoes cooling and enters a gas–liquid separator to separate the water, after which it proceeds to a molecular sieve dehydration tower for drying purposes.

(3) Cryogenic Adsorption Unit

The crude helium is first pressurized and then purified at low temperatures, with the cooling provided by liquid nitrogen produced in the crude helium extraction unit.

The crude helium is cooled to approximately − 180 °C by liquid nitrogen, then enters a separator for gas–liquid separation. The gas phase enters a cryogenic adsorber, where activated carbon is used to adsorb its impurities, with two sets of adsorbers used alternately. Through Cryogenic adsorption, all oxygen, neon, argon, and other impurities removed. As a result, gaseous helium with a concentration of 99.999% is obtained.

(4) Helium Liquefaction Unit

The pure helium from the crude helium refining unit, with a concentration as high as 99.999%, is further cooled and liquefied. The helium liquefaction capacity of this unit is 140 L/h. In the liquefaction process, cooling is achieved by using liquid nitrogen and helium cycle refrigeration (pressurization → two-stage expansion → heat exchange → pressurization and recirculation). The resulting liquid helium product is stored in a 10 m^3 liquid helium tank container and then filled into mobile dewars.

9.2.3 BOG Helium Extraction Plant of Yulin Yuanheng Energy Co., Ltd.

The BOG Helium Extraction Plant of Yulin Yuanheng Energy Co., Ltd. is built alongside an existing LNG plant. The first phase of the LNG plant has a liquefaction processing capacity of 1×10^6 m^3/d and includes one 2×10^4 m^3 storage tank. The second phase adds a new 1×10^6 m^3/d liquefaction unit and another 2×10^4 m^3 storage tank (Fig. 9.12). In the first phase, the natural gas has a helium content of about 300 mL/m^3 (ppmV), and the BOG has a helium content of 1%–2%. The helium extraction unit uses PSA, membrane separation, catalytic dehydrogenation, and low-temperature refining processes. In the second phase, the natural gas has a helium content of about 400 mL/m^3 (ppmV), with the liquefaction process adding two-stage denitrification flash tanks. The BOG volume is about 60 m^3/h with a helium content of 20%–30%. It uses membrane separation, catalytic dehydrogenation, and low-temperature refining processes, yielding approximately 200 cubic meters of helium per day.

1. Overall Process Flow

The BOG feed gas first passes through the PSA unit, increasing the helium content from (1–2)% to 30%. In this unit, two stages of membrane separation are carried

Fig. 9.12 BOG helium extraction plant of Yulin Yuanheng Energy Co., Ltd.

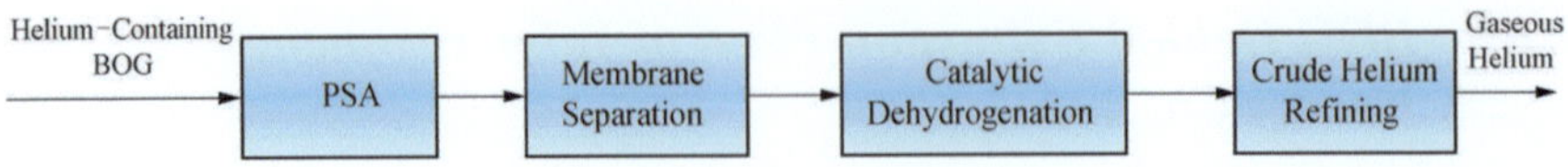

Fig. 9.13 Process flow diagram of BOG Helium Extraction Project, Yulin Yuanheng Energy Co., Ltd

out, producing crude helium with a helium content of about 96%. The crude helium enters the catalytic dehydrogenation unit, where hydrogen reacts fully with oxygen to form water, which is removed by molecular sieves. The low-temperature refining unit uses high-pressure, low-temperature adsorption to remove a tiny amount of residual oxygen, neon, and argon from the gas. As a result, 99.999% pure helium is obtained. The overall process flow is shown in Fig. 9.13.

2. Main Process Units

(1) PSA Unit

The BOG from upstream enters the PSA unit, where most of the methane and part of the nitrogen are removed, producing crude helium that enters the downstream membrane separation unit. The adsorption tower undergoes pressure equalization and desorption, with desorbed gas entering the inlet of the original liquefaction plant's BOG pressurization unit.

(2) Membrane Separation Unit

The crude helium is heated to 50 °C by an electric heater before entering the membrane separation unit, which consists of two stages. The crude helium first undergoes one-stage membrane separation, with the permeate from the first stage entering the second stage for further separation. The two-stage permeate has a helium content of 90%–96%. The non-permeate gas from membrane separation undergoes pressure regulation and then enters the inlet of the original plant's BOG pressurization unit.

(3) Catalytic Dehydrogenation Unit

Helium from the membrane separation unit enters the catalytic dehydrogenation unit, using two-stage catalytic oxidation dehydrogenation. To avoid overheating (less than 200 °C) that could be caused by fluctuations in hydrogen content, a stream of high-purity circulating helium is added at the catalytic dehydrogenation unit's inlet to ensure the hydrogen content in the feed gas entering the catalytic dehydrogenation unit is below 2%. After catalytic oxidation dehydrogenation, the helium (92%–98%) is dried using molecular sieves and then enters the low-temperature refining unit for further purification. Inside the dehydrogenation skid, the hydrogen content is monitored at the inlet using an online hydrogen analyzer. Hydrogen and oxygen analyzers are installed at the outlet to ensure the final dehydrogenation passes muster.

(4) Cryogenic Adsorption Unit

The helium gas, after undergoing catalytic dehydrogenation, is pressurized to 20 MPaG, and then enters the cryogenic purification unit. At liquid nitrogen temperatures, residual impurities are adsorbed. As a result, the helium purity is elevated to 99.999%.

References

1. Daly JW, Wayne J. Helium recovery from LNG. In: International petroleum technology conference; 2005.
2. McElroy L, Xiao G, Weh R, et al. A case study of helium recovery from Australian natural gas. Case Stud Chem Environ Eng. 2022;5:1–7.
3. Al-Harbi M. Helium 2-a new dimension to energy resources. In: International petroleum technology conference. Doha; 2014.
4. Fauve E, Grabie V, Grillot D, et al. Ras Laffan helium recovery unit 2. American Institute of Physics AIP. 2012; 1434:102–109.
5. Lindemann U, Boeck S, Blum L, et al. Turnkey helium purification and liquefaction plant for Darwin, Australia. AIP Conf Proc. 2010;1218(1):271–274.
6. Glowacki BA, Nuttall WJ, Clarke RH. Beyond the helium conundrum. IEEE Trans Appl Supercond. 2013;23(2):0500113.
7. Jones JK, Stacey JM. The recovery and liquefaction of helium from natural gas in Poland. Cryogenics. 1974;14(4):198–202.
8. Li JF, He LL, Chai LH. Present situation and suggestion of helium extraction from natural gas. Chem Eng Oil Gas. 2018;47(4):41–44 (in Chinese).

Correction to: Helium Engineering Technologies

Liming Pu

Correction to:
L. Pu (ed.), *Helium Engineering Technologies*, https://doi.org/10.1007/978-981-96-4086-7

The original version of this book was inadvertently published with minor errors in Chapters 2, 6, 7 and 8 which have now been corrected. The book and the chapters have been updated with the changes.

The updated version of this book can be found at https://doi.org/10.1007/978-981-96-4086-7

L. Pu (ed.), *Helium Engineering Technologies*,
https://doi.org/10.1007/978-981-96-4086-7_10

MIX
Papier aus verantwortungsvollen Quellen
Paper from responsible sources
FSC® C105338

If you have any concerns about our products,
you can contact us on
ProductSafety@springernature.com

In case Publisher is established outside the EU,
the EU authorized representative is:
Springer Nature Customer Service Center GmbH
Europaplatz 3, 69115 Heidelberg, Germany

Printed by Libri Plureos GmbH
in Hamburg, Germany